PERGAMON INTERNATIONAL LIBRARY
of Science, Technology, Engineering and Social Studies

*The 1000-volume original paperback library in aid of education,
industrial training and the enjoyment of leisure*

Publisher: Robert Maxwell, M.C.

ENERGY FROM THE WAVES

Second Edition
Revised and Enlarged

WITHDRAWN

THE PERGAMON TEXTBOOK
INSPECTION COPY SERVICE

Also of Interest

Pergamon Related Journals

ENERGY FROM THE WAVES

*The first-ever book
on a revolution in technology*

by

DAVID ROSS

Second Edition
Revised and Enlarged

PERGAMON PRESS

OXFORD · NEW YORK · TORONTO · SYDNEY · PARIS · FRANKFURT

U.K.	Pergamon Press Ltd., Headington Hill Hall, Oxford OX3 0BW, England
U.S.A.	Pergamon Press Inc., Maxwell House, Fairview Park, Elmsford, New York 10523, U.S.A.
CANADA	Pergamon Press Canada Ltd., Suite 104, 150 Consumers Rd., Willowdale, Ontario M2J 1P9, Canada
AUSTRALIA	Pergamon Press (Aust.) Pty. Ltd., P.O. Box 544, Potts Point, N.S.W. 2011, Australia
FRANCE	Pergamon Press SARL, 24 rue des Ecoles, 75240 Paris, Cedex 05, France
FEDERAL REPUBLIC OF GERMANY	Pergamon Press GmbH, 6242 Kronberg-Taunus, Hammerweg 6, Federal Republic of Germany

First edition 1979

Second edition 1981

British Library Cataloguing in Publication Data

Ross, David, *b. 1925*
Energy from the waves. - 2nd ed.
revised and enlarged.
1. Ocean wave power - Great Britain
I. Title
621.312'134   TK1457   80-41076

ISBN 0-08-026715-7 Hard cover
ISBN 0-08-026716-5 Flexicover

Printed in Great Britain by A. Wheaton & Co. Ltd., Exeter

Preface

This book is the result of a study lasting 30 months during which I have been educated, with great patience, by the leading experts in a completely new branch of technology. It has been written primarily for the non-technical reader who, in my opinion, needs to understand the energy revolution that is taking place around us. It concerns everyone who is interested in our future environment. The reader who finds technical matter difficult may find it easiest to skim Chapters 2 and 4 and return to them later; they are essential to an understanding of wave energy, but may be more easily absorbed in detail after the story of the discovery of wave power.

This is the first-ever book on wave energy and, I believe, the first time that a programme of scientific and engineering research and development has been reported in every detail, for the general public, while it was happening. The exciting and, at times, hilarious story of what happened behind the scenes in the Department of Energy in Chapter 5 shows the Civil Service in its best light and may, I hope, encourage those members of the Government who believe that an enlightened information policy would be more helpful in presenting our Whitehall machinery in a favourable light than the obsession with secrets which others prefer. I am particularly indebted to Dr. Freddie Clarke, Gordon Goodwin, Roger Pope and Derek Todd of the Department.

I have received great help throughout from Clive Grove-Palmer, the Secretary of the Wave Energy Steering Committee, who has been to me, and to the whole programme, a steadying influence. I am also grateful to Laurence Draper of the Institute of Oceanographic Sciences who has educated everyone in the wave energy field; Ian Glendenning of the Central Electricity Generating Board; Walton Bott of the Crown Agents;

Stephen Salter and his team at Edinburgh University; the Hydraulics Research Station; Sir Christopher Cockerell and his colleagues; and the gifted group at the National Engineering Laboratory. Some of the experts named in the book have read parts of it and have made helpful suggestions. The responsibility for what appears is entirely mine.

I am also endlessly grateful to Tamara, my wife, who has lived for two-and-a-half years with the waves roaring through not only our living room but also through the bedroom which is now the wave energy room and the television room which has had to become the bedroom, and who has borne with it stoically although her own country is Switzerland, one of the few in the world which has nothing to gain from the story that follows.

July, 1978

Preface to Second Edition

The book has been completely revised and brought up to date with a survey of all the main technical and political developments in wave energy during the past two years. The Introduction, the Addendum and the Conclusions have been re-written. However, I have left largely unchanged the historical record of the thinking and research as it existed in July, 1978, including the descriptions in The Cast of the four "first generation" devices. This is how they were at that time. The reader will discover in the new Addendum that they have been improved and that a "second generation" of inventions has joined them. But the first four deserve particular recognition as the pioneers in what I believe to be one of the most significant events in the history of the world's search for a benign, eternal source of energy to replace our diminishing stocks of fossil fuel and uranium.

May, 1980

Contents

The Cast

The Main Set Pieces

The Raft   Sir Christopher Cockerell has invented a trio of articulated pontoons which follow the contour of the waves. As the front one bobs up and down freely, and the second moves at a different phase, while the third stays relatively stable, the hinges swing backwards and forwards driving hydraulic jacks (pistons inside cylinders) which pump out hydraulic fluid and drive a motor. The motor turns a generator and the electricity flows.

The Duck   Stephen Salter, an engineer at Edinburgh University, has a line of bobbing vanes, shaped like cones and held alongside one another on a spine. They are known as Salter's Ducks because the beak nods up and down while the rump is designed to displace as little water as possible, enabling almost all the energy to be captured. The motion of the beaks drives rotary pumps which, in their turn, power a generator.

The Masuda   The inventor is a Japanese naval commander. It is an
Column   upturned canister with an air bubble above the water line and a hole in the top. As the waves rise and fall inside the canister, air is pushed out and sucked in through the hole. The motion drives an air turbine which is linked to a generator. The design has been improved and is being developed at the National Engineering Laboratory at East Kilbride, near Glasgow, where it is known as the Oscillating Water Column.

The Russell *Lock*	This is a power station standing on the seabed, unlike the other devices which float. It has been invented by the Hydraulics Research Station at Wallingford, Oxfordshire, where it is known officially as the HRS Rectifier (because it "rectifies" the direction of the flow of water). I prefer to call it after the Director, Robert Russell, and to use the more familiar word of Lock. It looks like a tall block of flats with lines of non-return valves facing the waves, like parallel rows of letter boxes. The water is forced into the monster and trapped in a high-level reservoir. The only exit is through a turbine into a low-level reservoir, rather like the system which rectifies the level in a canal lock. As the water flows down, it drives a turbine and, once more, produces electricity.

There are also plans being developed by Vickers Offshore, Belfast University, Lancaster University, Imperial College and many other centres.

The Main Characters

The Department of Energy
The Wave Energy Steering Committee at Harwell
The Central Electricity Generating Board
The Institute of Oceanographic Sciences at Wormley, Surrey
The National Engineering Laboratory at East Kilbride, near Glasgow
The Hydraulics Research Station at Wallingford, Oxon
The Crown Agents

And many others

Introduction

On the afternoon of April 29, 1976, the British Government announced that it was to spend just over £1 million on a two-year feasibility study into wave energy. The announcement was left to the Under-Secretary of State for Energy, Mr. Alex Eadie, and at a subsequent press conference in the Department every effort was made to play down its significance. One reporter was injudicious enough to ask whether it would be safer than nuclear power. The chairman, Dr. Walter Marshall, then chief scientific adviser to the Minister, stood up, looming and ominous, and replied sternly that he did not accept that nuclear power was dangerous. The spokesmen fended off suggestions from several of us that the amount of money allocated was insufficient. Yet within a year, on April 5, 1977, the Government tacitly admitted that it had been too parsimonious and the figure was more than doubled, to £2.5 million. Once again, the announcement was left to Mr. Eadie. The third "drop in the ocean", as one commentator called it, was £2.9 million as part of a package for all alternative sources of energy and this time the Minister, Mr. Tony Benn, did make the announcement himself in the Commons on June 6, 1978, defending the figure by claiming that "the limitation to making faster progress is not the level of funding but the state of the technologies involved".

Little had been done by the Government to publicise its efforts. There was a strong tendency to play down the significance of a development which, in the words of the first announcement, offers "a varying and partly unpredictable supply of electricity". The civil service knows how to throw a bucket of cold water over a new idea, particularly if it challenges entrenched interests.

Yet we are on the verge of a development as important as the discovery of steam power. The amount of energy available on what have been identified as the "likely wave power sites" off the Hebrides, the Scilly

xi

Isles, the mouth of the Bristol Channel and NE England has been estimated as "nearly five times the annual average demand on the Central Electricity Generating Board". That estimate does not come from some ardent propagandist, trying to get his or her hands on Government money. It comes from the very group that will have some awkward explaining to do if the lights go out and the factories come to a halt, the CEGB itself.

A team of its scientists and engineers concluded that the "development potential" along the British west coast "exceeds the present installed capacity of the CEGB". The figure was put at 120 GW – that is, 120,000 MW – compared with the CEGB's installed capacity of 60 GW and the average demand of less than 30 GW. The figures need to be treated with caution, for reasons that are explained fully in later chapters. At the time of writing in May, 1980, sympathetic observers were cautiously scaling down the figure and talking of the possibility of obtaining "only" one-third of our electricity from the waves, which would still be an enormous contribution. However, it can equally well be argued that the resource could be multiplied many times by installing parallel lines of wave energy generators at spaces of 160 km right out into the Atlantic. At this point, everyone is playing with numbers and the safest attitude is to fall back on the survey made by the Chairman of the CEGB, Mr. Glyn England. Addressing the staff of Fawley oil-fired power station, near Southampton, on July 4, 1978, he declared that the waves could "supply the whole of Britain with electricity at the present rate of consumption". He emphasised that he was talking about what "could actually be got to the electricity consumer". The speech went largely unreported and its significance has still not been grasped by informed opinion.

Mr. England based his statement on the research work done by his own people. I have been told that a policy announcement of this nature is considered, paragraph by paragraph, by all the experts concerned at the CEGB. The statement was in no sense an off-the-cuff remark. The text, which was supplied to me by the CEGB, contained evidence in the typing of cautious, last-minute revision.

He began by reiterating the CEGB view that nuclear energy was safe, reliable and efficient and appeared to be the most economic. "However, further upheavals in the energy scene are bound to occur" – meaning that a nasty accident or the increasing awareness of the long-term dangers of radiation and the pressure of public opinion (to which the CEGB and the Department of Energy are extremely sensitive) might cause an "upheaval".

It is also likely that some people in the CEGB are more conscious than the public of the fact that nuclear energy is inflexible (you cannot bring it in hurriedly to meet an unexpected peak in demand), unreliable (witness Dungeness), expensive (when the cost of research is included) and inefficient. But for the public, a brave front is maintained.

Mr. England continued that if an "upheaval" occurred, some of the renewable sources would be "more attractive".

He said that solar energy could be used for domestic water heating but would be prohibitively expensive for "electricity production", at least in Britain; apart from the shortage of sunshine, it was most scarce when we most needed it, in winter. In addition, a 2,000 MW power station such as Fawley would need an area of 50 square miles covered with solar cells, and the power station itself would still be needed on a dull day.

He dismissed geo-thermal energy as "speculative", though perhaps useful for providing hot water for nearby factories, homes and greenhouses. (The CEGB itself has since done considerable research into geothermal sources, but it remained a secondary source.)

As to wind power, it would need 4,000 aero generators, with a rotor 50 m across, to match the output of a single large power station. It would be possible, because of the lack of sufficiently-exposed sites, to meet only about 5% of our electricity demand in this way. He also reminded his listeners that they were not talking about picturesque, old-fashioned windmills but of "gaunt, massive structures".

Mr. England scouted the possibility of stationing aero generators in the sea and he suggested that this might produce a quarter of the CEGB's output. Since his speech, a great deal of work has been done on this suggestion but no-one has explained why the problems of building structures able, as he put it, "to withstand gales and heavy seas" (which are not always in phase) should be undertaken to capture the wind, when the waves are a concentrated form of wind energy.

He dismissed the idea of a Severn tidal barrage on the grounds that it could contribute "not more than 3% of the country's present total primary energy requirements", that it would be costly and could have a detrimental effect on the environment.

And so to wave energy: "As we see it at present, wave power is the most promising of the renewable energy sources. . . . Averaged over the year, there is about 80 kilowatts of power in each metre of wave-front approaching Britain from the N. Atlantic. This implies a total annual availability of 120,000 MW (120 GW) of power along Britain's Atlantic coast. However, not all this power could be harnessed. There would inevitably be substantial losses in conversion, and transmission losses, too, so that probably only about one-third could actually be got to the electricity consumer. Nevertheless, this is still a substantial amount of power — enough, in fact, to supply the whole of Britain with electricity at the present rate of consumption."

These remarks, coming from the man responsible for supplying the juice, should have galvanised the Department of Energy into an awareness of what is happening in the world of technology. Instead, it allowed events to overtake discovery.

The word "galvanise" comes from the name of an Italian physiologist, Luigi Galvani, and it is defined as "a term applied to the method of alleviation of pain and the cure of disease by means of a current of electricity".

Mr. England's statement had come one month after Mr. Benn's announcement of the third Government investment in wave energy. But it did not result in any change in the leisurely pace of the British programme. No extra money was produced. It left the country still gripped by what has been called "the bathtub syndrome". Tiny models were floated in laboratory tanks. Funds were provided for 1/10th-scale trials on Loch Ness and in the Solent. But all the research in the world will not provide the answers; the open sea always has a trick up its sleeve and it will be only when full-scale prototypes are launched that the answers will become known. The questions can be summed up: How do you collect the energy? How do you process it? How do you deliver it? How much is left for the consumer? And, what is for the CEGB the key question, how much will it cost? In reporting the facts, I shall not evade the difficulties. I hope also to demonstrate that they can be overcome.

We are at a fascinating moment in wave energy research when practically everyone, from the interested layman to the highly-qualified engineer who has taken an interest in the subject during the four years that it has been regarded as a practical possibility, has been going through a similar crest-to-trough syndrome. One starts by thinking: what a marvellous idea. We are surrounded by waves. They come in free. Why hasn't someone done something before? The sea is almost limitless. And the waves go on for ever. All we need is a water wheel and a cable. And the sea is at its roughest, and most productive, in winter, just when we most

need power. We can obtain electricity without pollution, without burning
coal or oil or uranium, without cost, without ever running out of the
"fuel".

Then comes the moment of despair. The cost of the devices will be
enormous. The sea will destroy everything exposed to its fury.
Maintenance will be at best expensive and at worst impossible. If we
absorb the sea's energy and make the surface calm, we shall silt up the
beaches and harbours. And how shall we manage when the sea turns calm
unexpectedly, as even the North Atlantic does at times, and there is a
surge of demand and a call to the power stations to meet it within
minutes?

And then comes the third, balanced stage – the zero crossing point, as it
would be called in wave terminology.

On the question of cost, it is possible to build up wave energy devices in
what have been called "penny pieces". That is, they are modular. Unlike
coal mines, tidal barrages and power stations, we do not have to invest an
enormous amount of money in one go. As to the power of the sea to
destroy them, we have built harbours and oil platforms. The Victorians
were able to build piers and harbours which still resist everything the sea
can throw. And maintenance will certainly be less than the cost of
ensuring constant supplies from coal mines. There *will* be problems – but
we have overcome them in the North Sea oilfields. The environmental
consequences do need to be examined but if the devices are far enough
away from the coast then the probability is that the loss of wave energy
will take perhaps centuries to make any significant mark. And the final
point, continuity of supply, is easily answered: no-one is suggesting that
we should abandon every method of producing energy and rely on the
waves alone. We can also extend the use of pumped storage and feed the
reservoirs from the sea and make the waves a firm source. The devices will
be stationed around different coastlines and it is almost never calm at the
same time on all the seas.

The more one examines the prospect, the more aware one becomes of
the difficulties, of the answers that already exist and of the prospect that
faces us of being the first power in the world to capture the energy of the
sea on a big scale – if we decide to invest in the project. It will at the least
enable us to stop using up our fossil fuels at anything like the present rate
to produce electricity. It will help to end the stupidity of burning oil and
coal and gas and uranium for the purpose. If will reduce our dependence

on nuclear power. It will provide us with an export trade in devices or patents which other countries will scramble to acquire.

And above all it will offer a serious prospect of continuing to maintain a civilised society based on industrial production without the certainty that one day, perhaps in our lifetimes, the fossil fuels will either run out or, if we are lucky, become scarce and expensive.

In this atmosphere, a great turmoil is developing among the people concerned, a "wind sea" as they might call it. There are rivalries among the contenders for producing the most favoured device. There are whispers circulating about the incompetence of the chap with the other invention. There are pressures from different interests − those, for instance, who want the devices close to shore or near to industrial centres; those who scorn the concept of stationary buildings on the seabed and those who deride the complications of a floating mechanical device.

What of the political background? Mr. Tony Benn was Energy Secretary when the wave power programme was launched in April, 1976 and he stayed in the background, unusually silent. He was probably encouraged in his discretion by his advisers who, understandably, were hesitant about being the ones who would submit a paper saying, "We ought to go abundance on wave energy". I sympathise. It is not an easy recommendation to make if one is in an official position and with North Sea oil available there was little sense of crisis.

Mr. Benn himself, it was whispered in Whitehall, had additional reasons for tip-toeing. He was strongly committed to an increase in the coal-burn (just as his Conservative successors were committed to decreasing their dependence on the miners). But Mr. Benn, fresh from the Department of Trade and Industry, was haunted by a word: Concorde.

He had been closely associated with the project and recalled how expenditure mounted and the combination of legal threats from the French, the need for employment in Bristol − his own constituency area − and the popular enthusiasm for a prestige ornament made it impossible to pull back when the growing financial catastrophe was blatant. Wave energy could be, from the public viewpoint, even more difficult. Everyone in Britain loves the sea. A nation of islanders has grown up aware that it has saved us from invasion and supplied us with food. The naval tradition, Royal and Marine, is strong. All of us have stood by the seashore and watched the waves roaring on to the beach in a useless expense of sparkling, wasted energy. Once the idea of wave energy captured the public imagination, it would be unstoppable. And that would mean that Mr. Benn was going to have to write out a cheque for major development — perhaps £1,000 million or at the least £250 million. It would still be a great deal less than the power stations, conventional or nuclear, the Severn barrage and the coal mines that are being built or contemplated.

For Mr. Benn there was, however, a stronger motive that should have drawn him to wave power.

One of Mr. Benn's heroes has always been President Franklin Delano Roosevelt, one of whose earliest achievements was to decide to spend his way out of the 1930s depression. He created the Tennessee Valley Authority. It was a time when there was, as now, a downward drift in expenditure and world markets were seizing up. The American Deep South was stagnant. The soil was barren after settlers had exploited the land without thought for the future. Woods had been cut down and deforrestation was creating floods.

Under the TVA, dams were built and lakes created. There was soil regeneration, national parks were designed and trees planted to soak up the water. And there was, above all, generation of cheap electrical power from hydro-electric schemes. The parallel is exact.

The scheme started in 1933, and by 1948 a total of 800 million dollars had been spent — a detail by comparison with today's prices and today's scale of US Government expenditure. No-one regrets or even remembers the cost. It was, as the Americans say, only money.

But it meant that three million people saw their incomes increased from below the poverty line and the surge of buying power helped to drag the US out of depression and indeed out of the appalling downward drift of self-confidence which could have reduced it to a second-rate power. Another Portugal or Holland, both of whom once also looked rich and impregnable. Or another Britain, whose decline over the past 30 years has been steady.

The Labour Government, which might have been expected to grasp an opportunity of spending money to head off a slump, and to spend it in a productive, useful way which would ensure a good return on capital, did not rise to the challenge. It was replaced by a Conservative Government committed to expanding nuclear power and cutting public expenditure. For a time, there was concern about whether investment in any of the alternative sources of energy would continue. The Conservatives ignored the Benn plan for announcements every spring of an annual programme.

But, by one means or another, the money (such as it was) continued to trickle through at a reduced rate, with serious cutbacks in some areas. The civil servants in the Department of Energy, and the Wave Energy Steering Committee at Harwell, managed to save the programme. But there was no indication of the massive, imaginative programme that could change the economic fortunes of the country by investing in our energy future and building up a world lead that would create new export markets.

There is little in the present industrial scene to persuade anyone that our fortunes will change. There is a consensus in believing that unemployment will continue at much the present level, or even increase. And one of the pillars of our past prosperity, the shipbuilding industry, is dying. Shipyards are without orders or surviving on Government subsidies which provide temporary relief to enable us to give away ships to other countries. Among the worst areas of unemployment are those regions, such as the Clyde, the North-East and Belfast where ocean liners were once built. The skills and facilities are still there — just. These are precisely the techniques and equipment that will be needed to build the structures, the size of giant oil tankers, that can produce energy from the waves.

But supposing, just supposing, says the ultimate pessimist, the whole idea fails? It is a fair question.

To me, it is inconceivable that, with the talent that has been devoted to wave energy research over the past four years, we could produce structures that will be complete failures. All the engineers, and all the scientists, just can't all be wrong. How much energy they will produce, how expensive it may be — these are the areas that can and should be discussed. But they are marginal. We need to develop a new way of thinking about energy.

We are not going to be in a position indefinitely to talk about which form of fuel we prefer. We now know that, sooner or later, the oil and coal will be used up. We must find a new, unconventional source of energy. And added to that we have the political message: we must find jobs, and they must be productive and help to create new wealth by helping industry to continue manufacturing the goods that we and the world demand.

We can give temporary lifts to the economy by cutting taxes and enabling people to buy transistors, electronic toys, package holidays and cars which eat up petrol faster than parasites can eat into the stonework of a cathedral. Or we can use the money for productive investment. Britain is facing a decision as significant as Roosevelt encountered in 1933. Neither of the two major political parties has recognised, and risen to, the challenge. But for the public, and particularly for the trade unions and the employers in shipbuilding, steel, electricity equipment and construction, there is now the chance to awaken to the importance of a new technical prospect as significant as that other British invention, the capture of energy from steam, and thereby revive dying industries and an economy which has grown over past centuries thanks to our conquest of the sea.

CHAPTER 1

How it All Started

"The motion and successive inequality of waves which, after having been elevated like mountains fall away in the following instant, take into their motion all bodies which float on them. The enormous mass of a ship of the line, which no other known force is capable of lifting, responds to the slightest wave motions. If for a moment one imagines this vessel to be suspended from the end of a lever, one has conceived the idea of the most powerful machine which has ever existed . . . ."

The words were written originally in French, in a style which echoed the majesty of the project: "La mobilité et l'inégalité successive des vagues, après s'être élevés comme des montagnes. . . ." Those words were filed on July 12, 1799 as the first-ever patent for a wave energy device by a father and son named Girard in Paris. Nothing is known of the outcome, if any, of the project. The idea was to build a gigantic lever, with its fulcrum on the shore and with a "body", the ship of the line as the Girard inventors called it, floating on the sea. As the body rose and fell "to a greater or lesser height according to the magnitude of the waves", the lever would work up and down and "could be applied to pumps, to bucket wheels, etc., or directly to mills, fulling machines, tilt hammers, saws, etc.". Attached to the application were a series of beautifully-drawn designs showing the floating vessel as a pontoon or flat boat. It was an invention similar to the project on which Sir Christopher Cockerell is working today on the Solent — an articulated raft.

The patent was discovered and translated by Alan E. Hidden, an engineer at Queen's University, Belfast, who is developing a novel turbine which is expected to make a major contribution. It was invented by Professor A.A. Wells, F.R.S., and is named after him. The beautifully-written translation is a reminder that engineers and scientists are literate and can even understand foreign languages, while arts graduates sometimes seem almost proud of being innumerate and innocent of what makes wheels go round.

1

The Girards did not vanish entirely. One can find references to a turbine which bears their name and which contributed to the development of the watermill. But it was to be a long time before engineering techniques and scientific understanding of the waves were to enable us to reach a point where wave energy became a practical possibility.

It was 174 years later, in the depressing winter of 1973, that Mrs. Margaret Salter turned on her husband and said: "Stop lying there looking sorry for yourself. Why don't you solve the energy crisis?" Stephen Salter, an engineer at Edinburgh University, had 'flu. The Arab–Israeli war had brought the oil crisis to a head and we had all become aware that energy was likely to prove at best expensive and possibly scarce. The Arabs might boycott the West; they would at least put up the price and create chaos with our economies. And, as so often happens, it was the increase in price that helped to concentrate the public's attention most wonderfully. When petrol went up from 30 pence a gallon towards a forecast level of £1, everyone suddenly realised that we were in trouble. The crisis was rather more than some doomwatcher crying out that the end was near.

Just how serious the long-term crisis really is, will provide material for numerous discussions until such time as the lights go out, if they ever do. Most futurologists are devoted to graphs which extrapolate a picture of the distant future from fairly recent events. They need not be taken too literally. We know that demographers are unable to prepare our schools for a rise or drop in the birth-rate even five years ahead. They tend to assume that last year's trend will continue indefinitely. We know that town planners can actually offer large sums of money to induce people to leave the inner cities because they are becoming over-crowded and then, within a few years, be appealing to them to come back. We know that the Treasury is regularly wrong in its calculations. And we must take into account the fact that most of the forecasts of our fuel consumption are based on the belief that growth will maintain itself at a steady 3.5%.

Let us assume that the forecasts are going to prove hopelessly wrong, in whichever direction you prefer. We are still left with the incontrovèrtible fact that we are using up our fossil fuels at a rate that cannot be sustained indefinitely. Even if there is nil growth, even if the Third World does not industrialise, the world will one day find itself without oil, coal, gas and uranium. And in the meantime by using up those fuels society will be polluting the air. And it will be condemning tens of thousands of human beings to spending their working life in uncivilised conditions underground, digging up coal. It is surely obvious that as this process continues, energy will become dear and scarce. Everything from pleasure motoring to the conveyor belts that bring the coal up from the mines will be threatened. At best, we will be forced to ration our use of energy or we can press forward heedlessly and leave it to our grandchildren to work out the consequences.

There is little doubt that society would choose the easy option and let the future take care of itself if the only consideration was a series of warnings. Happily, the increase in price has provided the missing element in making the crisis real.

It is important to understand the significance of this development. If there had not been this awareness of the situation, chiefly inside the Government but also backed by a public feeling of anxiety, then the stray remark by Mrs. Salter — if, indeed, she would ever have uttered it — would have had no fruitful outcome. Salter's Duck, if it had ever been invented, would have joined the long list of devices filed in the Patents Office in London, and similar ones in other countries, and that would have been that.

But, because of the price of oil in 1973, Mrs. Salter's suggestion may go down in history as being as significant as the day when, nearly 200 years ago, James Watt is reputed to have noticed the power of steam lifting the lid of a kettle and set off to invent the steam engine. Actually, this is disputed. There are authorities who claim that it is one of those nice fictions which has grown up, just like (in a similar area) there is no consensus about whether King Canute believed that the waves would obey his command. But there is no doubt that Mrs. Salter is the mother of modern wave energy devices. She is a psychologist, also at Edinburgh University, and she had rightly assessed her husband's reaction. He did not explode in fury as he nursed his wretchedness. Instead, he went off to his study and began work. His wife, he said later with a touch of sour humour, showed "callous indifference" to his misery. Instead, she gave him a precise objective: "What she wanted", Mr. Salter told me in his laboratory, "was something which would provide the vast amounts of energy needed, would be clean and safe, would work in winter in Scotland and would last for ever". He added drily: "It is a good thing for an engineer to have the design objective clearly specified." It is an even better thing to note that

(99)

349.

12 *juillet* 1799.

BREVET D'INVENTION DE QÙINZE ANS,

Pour divers moyens d'employer les vagues de la mer, comme moteurs,

Aux sieurs GIRARD père et fils, de Paris.

LA mobilité et l'inégalité successive des vagues, après s'être élevées comme des montagnes, s'affaissent l'instant après, entraînant dans leurs mouvemens tous les corps qui surnagent, quels que soient leur poids et leur volume. La masse énorme d'un vaisseau de ligne, qu'aucune puissance connue ne serait capable de soulever, obéit cependant au moindre mouvement de l'onde. Qu'on suppose un instant, par la pensée, ce vaisseau suspendu à l'extrémité d'un levier, et l'on concevra l'idée de la plus puissante machine qui ait jamais existé.

C'est principalement sur ce mouvement d'ascension et d'abaissement des vagues, qu'est fondée la théorie des nouvelles machines que nous proposons.

L'application en est aussi simple que l'idée première. Nous avons imaginé plusieurs moyens d'utiliser cette force; mais le moins compliqué de tous consiste à adapter ou à suspendre à l'extrémité

13 *

Figure 1.1   The first patent for a wave energy device, dating from July 12, 1799.

since that day Mr. Salter, and other engineers working on similar projects, have been doing successfully just what the good lady demanded.

To spell it out in a sentence: we could be on the verge of a new industrial revolution in which we shall be able to obtain all the energy we need from the ceaseless motion of the waves.

It is not a new idea. Since M. Girard and his son put their plan in writing, there have been more than 340 patents filed in Britain alone. Most of them, like the Girard scheme, are based on ideas that were not then, and in many cases are still not, practicable. But all of them have been inspired by what all of us have seen and felt. Anyone who has stood on the deck of a ship or a pier, or on a beach, must have been fascinated by the ceaseless, wasted surge of energy as the waves roll and break again and again in an apparently purposeless stream of excitement. We have all been brought up to know that the water wheel was a main source of power before the steam engine. It is logical to conclude that a water wheel, poised in the ocean, could do the same job. There is a snag: despite their appearance, the waves go up and down in a peculiar motion which we shall study in detail later. They do not move in the way that a millstream does, flowing down a hillside and driving a wheel. The tide does flow in that way but it has disadvantages. In Britain, there is only one area that is really attractive for a tidal barrage and that is the Severn. It has some disadvantages, particularly the huge capital cost. The attraction of wave energy is that it can be built up on a modular basis, in "penny pieces". And it is not a once-off enterprise.

I do not wish to decry tidal energy, which may be more suitable for some countries than it is for us. Equally, solar energy can be a major source of power in sunnier climates and even in Britain it can be a useful asset for, say, hot water in individual homes. Wind energy may be a major advantage in countries like Canada with many miles of uninhabited mountains but it would be wrong to think that it could ever be environmentally acceptable in Britain: the charming idea of windmills dotted around the countryside has no significance in terms of energy. It can, like solar energy, light up the odd farmhouse but we should need enormous, very noisy aero-generators, with horizontal arms sweeping across our beauty spots if we were ever to contemplate using the wind as a major source of power. The wind energy people are now, belatedly, thinking of putting their aero-generators out at sea. They missed the bus

Figure 1.2   Stephen Salter in his laboratory at Edinburgh University.

on that one, when wave energy won Government backing. And now it is difficult to see why one should go out to sea, build vast structures in the ocean and try to capture power from the wind when the waves are actually a concentrated form of wind power.

Just how do the waves work? The best description comes from Walton Bott, consulting engineer for the Crown Agents, who had done extensive work on capturing wave energy in Mauritius. In an address to the Royal Society of Arts, he said: "The visible effect can be seen from the motion of a floating bottle or ball which, as the wave passes, describes a circular motion returning to approximately its original position. I have never yet been able to think of a suitable mechanical analogy but perhaps the pushing by hand of a ridge in a carpet from the centre to a wall edge is somewhere near; for a wave does move across the carpet but every strand of pile remains firmly in position."

It is almost certain that the problem of capturing this energy could not have been tackled successfully before now. Try straightening out a heavy carpet and you will begin to understand the problem. But that is only the beginning. When you have expended your own energy on straightening the carpet, you have to start to think about capturing the product and converting it into electricity. You have the problem of building a device the size of a giant oil tanker and structuring it so that it will absorb and process the energy of the sea without being swept away or broken. It must be, as they say in industrial negotiations, firm but flexible.

And then you need the money. Who would have invested in such a scheme before the oil crisis? Society is content to allow cargoes of oil to be carried thousands of miles with occasional accidents and pollution of the beaches and of the sea. It is prepared to permit human beings to be dropped thousands of feet below the earth's surface to spend their working lives in darkness and in danger, digging at the coalface. It is prepared to use up our capital to provide our liquidity, to use fossil fuels to produce electricity.

Sir Christopher Cockerell, the inventor of the Hovercraft, is another engineer engaged in producing wave energy. He admits, with embarrassment, that he uses oil-fired central heating for his own home. And he added when we discussed the problem: "To use our oil to produce electricity is just terrible. Anyone who lives by using up his capital is a mug."

Sadly, the attitude of some of our environmentalists is not helpful. A distinguished scientist closely connected with the top echelons of the Labour Party remarked to me that he had been brought up in the "humanist" tradition, which was now lacking. He had been educated as a scientist at a time when it was standard belief that technology could do anything. All problems were soluble. All knowledge was knowable. Science held the key to the future. "Soviets plus electricity equals Socialism", as Lenin put it. Then came the backlash. A post-war generation has turned against science and engineering and has come to believe that pollution, destruction, noise, ugliness and — despite the optimistic forecasts of the scientists — human wretchedness, unemployment and poverty are the natural product of an industrialised society.

They have not only forecast catastrophe if we continue to rely on fossil fuels; they have sometimes given the impression that they would welcome a return to a pre-industrial condition, a Gandhian, village-based community. They have tended to discount the effectiveness of alternative sources of energy, except as a means of lighting and heating individual homes which could then cope without needing help from industry. They appear often to be possessed of a sense of guilt whenever a machine does a job that previously needed muscle.

Plainly, it would be a pleasanter world if aircraft and motor cars and factory chimneys did not disturb the peace. It would also be a great deal more uncomfortable, particularly for those who had to provide the manual labour to replace the machines. And it is not going to happen.

We have today sufficient technology to ensure that people will be able to continue to live with the help of gears and levers and turbines and man-made power. The problem to be solved is to produce that energy without using up the last of the fossil fuels and leaving our descendants to take the consequences. Electricity is the cleanest way of doing it, but not if we have to dig up coal and burn it, not if we have to exhaust our oil stocks. Our problem is to continue what we have come to regard as civilised or at least organised society without pollution, without depriving the future of that civilisation.

My favourite doomwatcher travels everywhere by bicycle. He was returning home up a steep hill when some schoolboys jeered. "Get off and push", they shouted. He pedalled on grimly, reached the top and went home with a rupture. The operation was performed in a hospital powered by electricity, generated by power stations using up the fossil fuels which my friend had sought to conserve.

The waves provide a better source of power. The waves go on for ever. They will continue to rush tirelessly, ceaselessly, against the shore, against the piers and breakwaters and, if we handle our resources properly, against the wave energy generators that can contain and process their fury to the advantage of us all.

CHAPTER 2

What is a Wave ?

When Stephen Salter began his studies late in 1973, he set out first to visit the Institute of Oceanographic Sciences (IOS), a bleak collection of buildings on a side road somewhere between Witley and Wormley in Surrey. Everyone involved in wave energy since then has followed the same path, seeking out Laurence Draper who has shown unlimited patience in spelling out the basic facts about the waves, a subject on which he is one of the foremost experts in the world, to the growing stream of inventors, experts in other disciplines such as mechanical, constructional and electrical engineering, and to me. Mr. Draper has the gift of simplicity. He needed it less for that first visitor as Salter is a physicist. But the waves are an esoteric branch of physics.

Salter's first need was information and then a suitable site for experiments to find out just how much wave energy there was available, and Mr. Draper knew how to find it. He telephoned to Rear Admiral D. A. Dunbar Nasmith, C.B., D.S.C., in Inverness where he is Deputy Chairman of the Highlands and Islands Development Board. It was not easy. It will be appreciated that there are problems in launching oneself into this topic. "You see, Admiral, there is a scheme to get energy from the waves and we would like to use your beaches . . . ."

Unknown to Mr. Draper, the Admiral had a visitor in his office at the time who was to prove a key figure. Father Calum MacLellan is the priest in charge of St. Mary's at Benbecula in the Outer Hebrides and his parish was to prove the most suitable site for the first tests. The sea in that area is beautifully rough and therefore productive; and a major industry is lobster fishing, which means that trawlers stay away and would not therefore be fouling the Waverider Buoy which Mr. Salter and Mr. Draper planned to tether off shore to obtain accurate measurements of the waves. Benbecula is the island where the Reformation stopped half way and the southern half is correspondingly fierce in its Catholicism. Without Father MacLellan, a foreigner has no hope.

But on that day, the only concern was to obtain the support of the Admiral, who gave it readily, and has remained a keen enthusiast. Mr. Draper, and a colleague, John Driver, who is an expert on wave recording instruments, set off for the Outer Hebrides. They were met by an Army colonel from the nearby rocket range and he, like practically everyone else who first encounters the idea, was intrigued. He took his own car and drove the two men over the area looking for a suitable site. Father MacLellan's parish was the answer. But how could one win the cooperation of the local villagers who were suspicious of any outsiders, let alone a crackpot pair of boffins who wanted to plant bits of machinery in the lobster beds?

"It was then that Father MacLellan entered the scene", Mr. Draper told me. "It is not too much to say that the villagers are devout Roman Catholics and Father MacLellan's word is law. He is an impressive man, a person with whom we rapidly developed a friendship based on respect and if we had not been able to reach agreement with him, it would not have been on. But at short notice he had interrupted his busy schedule and hurried over to the military base where we discovered that he was on good terms with the Admiral and so had heard about us before we arrived, and he said he felt sure that it was a worth-while project."

Father MacLellan confirmed this to me, but he is retaining a cautious attitude. "There is nothing wrong with putting down an instrument to find out whether wave energy is going to be a feasible proposition", he said. "And I do not believe it is contrary to God's purpose. But if it does become feasible, one would have to consider very carefully in case it changes the whole environment and interferes with the lobster fishermen. Our information was that they wanted one single buoy. There is no possible objection to that. But they wanted to be sure that lobster fishermen would not be picking it up and looking at it and that it might disappear. They have lost other buoys."

With the assurance of acceptance, the two men began the first scientific study of the waves close to the west of the Outer Hebrides.

How does a wave travel? It was, as in so many other areas, Leonardo da Vinci who was an early exponent. He noted that when the wind blew across a field of corn, it looked as though there were waves of corn running across the field. In fact, the individual heads of corn were making only a slight movement and when the wind dropped they were back almost where they had started. A more common description nowadays is to compare it with the movement in a skipping rope. You waggle one end of it and a wave *form* travels to the other end.

In his address to the Royal Society of Arts in March, 1975 (a generation ago, in wave energy terms) Mr. Bott, the first practical engineer to tackle wave energy in recent years said:

"Each water particle describes a circular path in a vertical plane at right angles to the wave line, the diameter of these orbits diminishing rapidly with water depth at right angles to the wave line, ultimately becoming zero."

It is essential to grasp the concept that the waves are not tides or currents. It has been possible to identify waves arriving in Cornwall as having originated in the Antarctic but this does not, of course, mean a current of icy water cutting across the Gulf Stream.

The seaman divides waves into two main types — those generated by a local wind and known as the wind sea, and those which come from remote storms and arrive in the form of a swell sea. In layman's terms, it is the difference between a busy sea of lively bobbing waves and a long, sweeping, rolling sea. These are Mr. Bott's water particles turning in a circle and then jumping about as they are squeezed. And then: "the wave rears up until it catches its feet, so to speak, on the ocean floor and falls head over heels as it crashes on a reef or beach".

Our knowledge of wave energy is nearly all recent. There were some studies over the past 150 years, but no real development until, as in so many other activities, the needs of warfare provided an incentive for government expenditure. It had been the same with steam power: the Napoleonic wars increased the cost of animal feedstuffs and made steam more economic than horses. It had been the same with nuclear power: Sir John Anderson, the Minister in charge, noted: "In four years, our scientists solved a problem that in peace might have taken 25 to 50 years" (referring to nuclear power). And with the waves, it was the need to invade continental Europe in 1944 which suddenly aroused interest and a readiness to spend money.

Fleets of landing craft had to be put ashore in Normandy. An artificial harbour called the Mulberry had to be towed across the Channel and installed. There was little satisfactory information available. I recall an appeal going out from the Ministry of Information for holiday snapshots showing the beaches of northern Europe. The Admiralty set up a Research Laboratory Group W (the mysterious letter stood for Waves but it was secret, see?).

That unit became the nucleus of the National Institute of Oceanography and has now been re-christened the IOS. It is run by the Natural Environment Research Council under the Department of Education and Science. The other organisation, about which more later, is the Hydraulics Research Station at Wallingford, in Oxfordshire, which is dedicated to checking the validity of engineering designs for harbours, ship moorings and so on. It has a plan of its own for wave energy, reflecting its involvement in the civil engineering side. The IOS covers basic research into the oceans, from the character of the waves, the behaviour of the water including tides, to marine biology, chemistry and the geology of the seabed.

The bespectacled physicist in charge of those early IOS studies on wave energy was Laurence Draper who somewhat improbably owes his involvement to an interest in rock climbing. He was a student at Nottingham University in the early fifties, after being a scientific assistant with ICI. He felt little inclination to return there and had not much idea of what he would do with his degree. But he was a keen mountaineer and one day read in a climbing magazine the obituary of a talented climber named P. J. H. Unna. Not long after, he was reading the scientific journal *Nature*, and came across an article on the waves by P. J. H. Unna. He thought that if one mountaineer could be fascinated by the waves, surely another could be, too. Later, he came across a report that the newly-formed NIO was measuring waves, so he wrote to them, to ask if they would like a helping hand in the holidays.

They had never before had a holiday relief but the idea appealed. He spent a summer with them and the director invited him to come back when he had his degree. He did just that. On this completely illogical, unscientific happening, linking the mountains and the sea, depends much of our future today.

Mr. Draper now looks back coolly at the primitive efforts of Research Laboratory Group W as they wrestled with the problem of crossing to Normandy. "They got it right in 1944", he told me, "but only just. They failed to fully understand the problem of swell." Some of the chaps in the landing craft would agree, feelingly. Trust the Navy, as we used to say in the RAF.

He is at the centre of what he happily calls "one of today's growth industries, the systematic collection of instrumentally-measured wave data". He and his colleagues are ahead of everyone in the world with knowledge of what is called "the wave climate". The Institute of Oceanographic Sciences' methods are used world wide. The Department of Energy is well aware of its importance and supports it. No other country has a map of predicted extreme waves right round the coast. That map provided the basis of all North Sea oil operations and was able to save the Americans from what, in Mr. Draper's opinion, would have been certain catastrophes. They were convinced that their experience in the Gulf of Mexico would serve them well in the North Sea. In particular, they were rather over-confident because they had the experience of coping with hurricanes. "But", as Mr. Draper told them, "hurricanes move very fast and so they do not produce the biggest waves. They asked us what would be the biggest that they should expect if the platforms were to be there for 50 years so we looked at the wave data and came up with a figure of 17 m. The Americans seemed to think that we were just trying to cover ourselves. They thought that 12 m would be the biggest. (Remember that the difference in cost then was about £4 million for 3.75 m on each platform). But they have since seen waves bigger than that themselves and the oil platforms would probably have fallen apart if they had followed their own instincts. Even so, there have had to be restrictions placed on the use of the early platforms, which have to be evacuated in rough weather, and which have had second platforms built alongside to take people off."

It is worth pausing to appreciate that a wave 17 m high is the size of, say, a six-storey block of flats. The waves off the Outer Hebrides, where wave energy generators will probably be based, can exceed 24 m, the "worst" sea in the world, probably worse even than Cape Horn, and therefore the best for our purposes.

The science of wave climatology has grown spasmodically, usually as a poor and almost unknown relation of other branches of science and

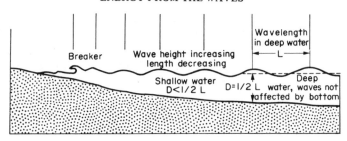

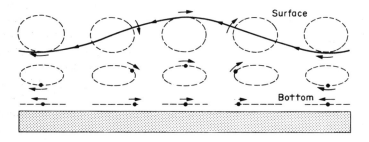

Figure 2.1   Cross section of wave profile showing how decrease of wave length is accompanied by increase in height and steepness as the shore line is approached. In deep water, the waves are not affected by the bottom. Then, as the bottom rises, the depth equals half the length. In shallow water, the depth is less than half the length, the wave height is increasing and the length decreasing and the water is squeezed between "sea level" and the shore. The sea level never rises; the seabed does.

engineering. It was needed for wartime landings and then came increasing demands for information for the building of lighthouses, pipelines, breakwaters, harbours and, in recent years, for oil platforms, hovercraft and hydrofoil services. And now comes wave energy, its needs arising to justify the collection of all this once-purposeless information, like feathers folding into a dove's tail.

The information has been built up with the aid of three main types of wave recorder, on the seabed, in the surface and on ships. It is available so that a civil engineer can collect a graph from which it is easy to read off the expected wave height in any area, provided certain basic and easily-obtainable information is available.

One of the first recording devices was the spark plug recorder. A series of spark plugs, each with a horizontal electrode, are stationed in a vertical line with the plugs a few inches apart. As the sea water rises and falls, the plugs under water are shorted and it is easy to collect a record. It is ingenious but not as sensitive as others which are designed to register a difference as small as 0.25 mm (1/100th of an inch). One such development is under-water measurement with a pressure sensor which has the advantage that it is less likely to be damaged by shipping. Its disadvantage is that it can be used only in fairly shallow water, down to about 12 m, because waves do not penetrate too far down into the sea. It is a metal box, sitting on the seabed, which records water pressure. As a wave crosses the site, the depth increases and so does the pressure, and the box records the extra height. The information is recorded on a graph or a cassette, either in the box itself or back on shore to where the information is relayed by cable.

A more sophisticated device, which can be used in deep water, is the Shipborne Wave Recorder which was invented as long ago as 1951 by an IOS scientist, Mr. M. J. Tucker. A pressure sensor is mounted inside a stationary ship, such as a weather or light ship, below the waterline. Then a hole is bored in the side. As the water deepens and shallows outside the ship, the pressure sensor records the changes. But it has to be coupled with an accelerometer which registers the movement of the ship itself. Then the two measurements are added together automatically and the graph prints out a picture of the waves on the site. This has been the main wave-recording workhorse of the IOS since the mid-1950s.

It was put first on one of the IOS research ships but they are required for many duties and are moving around all the time. The same applies to weather ships — for instance, the best known site, Station India (in the North Atlantic, despite her name) would be manned by half a dozen ships with turns of duty of four weeks on and two weeks off. Nevertheless, one of the Meteorological Office's vessels, *Weather Explorer,* was fitted with a recorder in 1953 and by 1965 she and her successors had amassed enough information at Station India for the wave climate there to be known quite well. In fact, these measurements have become the basic information on which the wave energy potential of the Atlantic has been calculated.

Subsequently, Trinity House gave permission for one of their Light Vessels to be used, but as they remain on station for three years at a time they have to make a special trip into dry dock to be fitted because, as Mr. Draper remarked, Trinity House is rather strict and doesn't like chaps drilling holes below the water line while ships are at sea. Meanwhile, a replacement ship has to be provided. That costs money. So, up to that point, it had all been done on a string-and-Scotchtape basis.

It was then that, once more, Defence came to the rescue. The military needed information about the waves for radar research. There was no shortage of money. A Light Vessel called the *Morecambe Bay*, stationed in the Irish Sea, was swiftly called in and a replacement sent out. Trinity House and the IOS worked fast and within six weeks had installed their recorder and the ship went back on station. The experiment was intended to last for a few months and, indeed, once the military had the information they needed, they were happy to let the IOS have their bits of equipment back. But the IOS were in no hurry. They had suddenly realised that for the first time they were receiving instrumentally-measured details of the wave climate at an off-shore location.

The main purpose of the exercise at that time was to enable the man in charge, Professor J. Darbyshire, to develop a method of forecasting the height of waves by what is known as "hindcasting". If you know the strength of the wind and the distance and length of time over which it has blown, you can calculate the height of the waves. But data had to be obtained before the method could be developed.

As soon as the IOS realised how valuable were its first consistent records from a permanent site, it approached Trinity House for more facilities. It obtained permission to remove the recorder from the Irish Sea and install it on Smith's Knoll Light Vessel in the southern North Sea in 1959. Another set of records was obtained. Then records were collected from the Helwick Light Vessel in the Bristol Channel and from the Sevenstones Light Vessel off Land's End. Eleven other Light Vessels were eventually employed, and several are still operational today.

The IOS did encounter one area that proved unexpectedly difficult. There is no Light Vessel in deep water off Scotland because, as Mr. Draper put it, "there are all those nice little islands to put lighthouses on". But he stumbled by chance on a reference to a Norwegian Rescue Vessel. The British had once been asked to contribute to the cost of it and had refused and everyone concerned from then on ignored its existence. Mr. Draper wrote to the Norwegians and explained what he was trying to do. And he was able to add that the information he obtained would almost certainly be of use to them. Happily, they proved to be less narrow in their views than the British and a recorder was installed in 1969 and proved to be a vital source which has contributed greatly to the safety of the platforms built in both the British and Norwegian sectors since then.

Subsequently, another recording instrument has been in increasing prominence. It is a Dutch invention called the Waverider Buoy, a bright yellow and orange sphere, 80 cm in diameter, moored to the seabed and tethered (literally) by a piece of elastic. There are dozens of them now functioning, one of them in the sea off Benbecula. They transmit their information over a radio link and their only disadvantage is that they can be damaged easily by shipping or by curious seamen. The aerial is particularly susceptible to light-fingered gentlemen.

The work of collecting data had started by a happy accident but it was to prove timely. The discovery of North Sea oil was round the corner and the oil companies were soon desperate for information. It was provided for them but it is worth noting that their gratitude has not taken the most practical form. Some of the companies still want their environmental data free, which drives the more enlightened ones who pay for their own studies to act as though they were afraid that a playback of information might help their rivals, and they treat their knowledge as a commercial secret. They were all beneficiaries of the sort of information that only governments would have subsidised in the days when its practical advantages were not apparent. But − business is business, once the assets have been gained at public expense.

The significance of the height of a wave is that it gives an accurate indication of its power. The wave consists of kinetic and potential energy − in simple language, that is the power in the falling water (potential) and the force moving forward (kinetic). Research has shown that both are roughly equal so the height is an excellent indication of how much energy it represents.

Another key factor is the frequency with which the waves arrive. This depends to some extent on the length of what is called "the fetch" − the distance over which the wind has been blowing over the sea before reaching the point of interest. It has to be a straight line without interruption from land. This, combined with knowledge of the wind speed, enables scientists to forecast wave heights by that method of hindcasting. Records exist going back 100 years of winds over the sea. By knowing the expected wind speed at any season, the distance of the fetch and the length of time that the wind has blown, it is possible for anyone with a minimum of technical knowledge to read-off on a chart the predicted wave height. One can also read-off another chart the wave period, the time it takes between the arrival of one crest and the next.

This period can vary from two seconds to above 20. And by reading backwards from the time interval between the arrival of waves of different period, it is possible to discover where the waves originated. Thus it is possible to say occasionally that a wave reaching Cornwall originated in Antarctica.

Sea waves are unlike radio waves which always travel at the same speed. In a severe storm, there will be waves with periods from 2 seconds up to say, 20. The longer the period between waves, the faster they travel. So waves with 20 seconds between them travel faster than those with 19 seconds. If you are a long way from the source of the storm which first whipped up the sea, then the waves of a 20 second period will arrive first, having outpaced the waves of 19 seconds, which will have outpaced those of 18 seconds, and so on. The further the waves have travelled, the greater will be the lead of the longer period wave over the shorter. It is like two planes taking off at the same time, one going at 300 km/hr and the other at 600 km/hr. Just after take-off, they are close together. As the distance gets greater, the gap between them grows.

If you know these facts, and the time difference between their arrivals at some other point, you can work out how far away they started. An example from the IOS charts will help to demonstrate this. Take a wind speed of 50 knots and a fetch of 100 km. Then what is known as the "significant wave height" will be 6 m. Assume the same windspeed of 50 knots and a fetch of 1,000 km, then the "significant wave height" will be 8 m. It is obvious to anyone who has ever done an algebraic equation that armed with any two of these three facts, one can find the third. But it took a long time and extraordinary talent to reach this point, which is going to prove vital to all our futures.

I have, deliberately, over-simplified the problem. How, for instance, do the waves in a storm vary so much in height? Mr. Draper's answer gives some indication of the difficulties that he and his colleagues and predecessors have had to grapple with:

"Each one of the waves was generated by the same wind over the same water and at the same time, and yet they come in sizes over a range of more than 10 to 1 in height. The explanation is that waves are remarkably short-lived creatures, no storm wave ever exists as an identifiable entity for more than about two minutes. Even the 80-foot monster has a very short moment of glory and if one could follow its progress it would be seen to diminish in size and within a couple of minutes it would subside into the random jumble of the sea, never again to reappear in that form. This behaviour is a consequence of the fact that wave energy in the sea is locked in a very wide range of wave components, each with its own height and period.

The key to the problem is that each component travels at a rate determined by its period, so that the faster components (which have the longer periods) will continually overtake the slower ones. Consider, for a moment, the simple case of a wave system consisting of two components travelling at slightly different speeds. As the crest of one component overtakes the other one, a bigger wave will temporarily appear. Similarly, as a crest overtakes the other component's trough the sea will be relatively placid. In a real sea there are not just two but millions of wave components (an infinite number if one takes it to its mathematical limit), each travelling along at its own pre-ordained speed. Just occasionally, purely by chance, a very large number of components will all be trying to overtake each other at one point in space and time, and the unfortunate mariner who happens to be there too will be able to report an enormous wave."

One begins to see the problem of identifying — the problem. That is one reason why the engineers and scientists grappling with wave energy are even today, when the need for more energy is starker than ever before, tending to caution. The most elaborate tests, in laboratories or lakes or sheltered coastal waters, can never reproduce the fury of the open sea.

Yet it is a fact that, even before they knew what they were facing, Victorian engineers were able to build harbours, breakwaters and piers which could survive the sea for 150 and more years. The surge wave which swept away part of Margate pier in January, 1978 was evidence not that the sea was irresistible but that what civil engineers could do in 1800, in providing a structure that would last until 1978, can be eminently better done now.

One must not be unfair: a pier, like an oil platform, is designed to allow the sea the minimum resistance while wave energy generators will need to absorb and process its energy. They will, therefore, be subject to greater stress. But technology has advanced a long, long way since those seaside piers, and even the North Sea oil platforms, were erected.

We shall not know for certain until the generators are on site. It may then well be that those of us on the shore, whose innocence is matched only by instinct, may well be more right than the experts in urging a swift leap ahead. There is evidence to support us.

For instance, I have used tι.? phrase "significant wave height" and it is defined by the scientists as "the average height of the highest one-third of the waves". This formula has been reached after studying the print-outs of wave recorders. And the extraordinary thing is that the most modern data, obtained by the sophisticated electronics that have been developed in the last 20 years, confirm the old mariners' stories about the waves they survived. As Mr. Draper puts it, "If the heights of 99 consecutive waves are measured, the significant height is the average height of the 33 biggest and it is this height which is very close to the figure which an experienced seaman would give if one were to ask him to estimate the wave height."

Thus, in 1839, a Captain Robert Fitz-Roy reported a wave 18 m high in the Atlantic and an American naval ship, the *Rampapo*, claimed to have survived a wave of 34 m in the Pacific in 1933. Fantasies, one might have said – until recently. But now we have an authenticated report of a 27 m wave hitting a drilling rig off Vancouver Island in the Pacific and in 1971 a wave of 26 m was experienced by the weather ship *Weather Reporter* out in the North Atlantic with a shipborne wave recorder to confirm it. So the old salts, with just their eyes to rely on, have had their tales confirmed by modern science. Instincts are not always wrong.

On the other side of the coin, Mr. Draper has a cold douche of water to throw over a hallowed myth. There are, he has noted, such things as "freak" waves which exist for a brief moment as countless wave components come together. They rarely come singly. Their usual appearance is of one huge wave followed by a few slightly smaller, stretching over a distance of hundreds of feet, "all following each other obediently in a regular procession along the Loch before disappearing as they get out of phase (or diving for food, if you prefer it)". And he adds, as a good scientist must: "It should be pointed out that such a plausible explanation cannot prove that the Loch Ness Monster *is* just a figment of its disciples' imaginations . . . ."

When he wrote that, in 1971 in *Motor Boat and Yachting*, he could not have guessed that within seven years there would be a major project, equipped with all the latest technological equipment, based on the shores of the Loch for the purpose of launching a string of wave energy devices, and with the prospect of discovering, as an accidental by-product of research designed for a different purpose, if there really is a monster around.

CHAPTER 3

Back to the Water Wheel

Wave energy means that we are going back to the water wheel. And we are doing so amid an uncanny echo of the problems that accompanied first the introduction of water power, including the "religious" turmoil, and then the objections which were raised to steam energy. The closest parallel is with the late 18th century when James Watt's steam engines were producing only 11 kW of energy and factories were being built for the textile industry driven by as much as 190 kW of water power. It would have needed considerable vision at that stage to accept that steam would soon be ousting water as the prime source of energy. The watermill had been the earliest form of mechanisation, freeing animals and slaves from the treadmill and peasants from grinding corn by hand. It occupies a prime place in everyone's awareness. It yielded reluctantly to steam and even today, in hydro-electric power, water continues to be a major source of electricity.

At the time of the Domesday Book, in 1086, there were 5,624 mills in operation in Britain, nearly all of them water mills, serving a population of only two million − one mill to every 400 people. Then came progress. Steam was captured and its appeal was irresistible. But *was* it progress?

Mr. David Braithwaite, architect and editor, sums up the position in his introduction to *Windmills and Watermills* by John Reynolds. "If it is true that the discovery of new sources of power has been the basis for the progress of civilisation, and the rate of progress has been determined by the amount of energy available to man, then the superseding of wind and water power must surely be celebrated" he writes. "But it is an inescapable fact that the harnessing of the chemical forces of fire and steam to drive engines and turbines has proved wasteful and has polluted unimaginably our atmosphere." Note that Mr. Braithwaite did not regard the using up of our fossil fuels as worth a mention. He was writing in 1970, far away in our awareness of the energy crisis.

The great promise of wave energy is that there is, for perhaps the first time since the water wheel was discovered, an opportunity of embracing the concept called progress without needing to lament its undesirable side-effects. The possible environmental consequences are all, so far as we know today, either favourable or soluble. The problems are: Can the waves ever provide our major source of energy? Which type or types of device should we choose? What will be the social consequences? And, to mention the almost-unmentionable, is it morally acceptable that we could actually use natural resources without penalty? This is the modern adaptation of the ancient Romans' feeling about water power, the "religious" argument brought up to date with a collective guilt complex added. It is, more simply, the feeling that there must be a snag somewhere.

The first water wheel, known as the Persian wheel, goes back to 200 B.C. It worked just as most people nowadays imagine, if they think about water wheels at all. That is to say, a wheel stood in a stream of water and the wheel turned as the water flowed. The wheel had a series of buckets hanging from its circumference, like seats on the Big Wheel in a funfair. The buckets scooped up the water and emptied it just before reaching the top, either on to another higher area which needed irrigation or into an aqueduct. It was another 100 years, in 85 B.C., before there was the first reference to wheels being used for a secondary, mechanical purpose.

A poem by Antipater of Thessalonica reads:

"Cease your work, ye maids, ye who laboured at the mill . . .
For Ceres has commanded the water nymphs to perform your task."

The mill was where the "maids" had worked at grinding grain between stones by hand. The first mills were of the pestle and mortar type, gradually evolving into what has come to be known as a saddle quern — the lower stone was ground into the shape of a saddle as the upper stone rubbed against it. A round stone, with a handle, was developed and donkeys were used to drag it. And then man's genius brought together the water wheel, driven by a natural force, and the millstone. Humanity had reached a bread and water economy.

There is no record of how Antipater's "maids" responded to the prospect of redundancy but when water wheels were next used to replace labour, in Rome, there were considerable perturbations. The wheel is referred to by Vitruvius in *De Architectura* and he mentions a toothed gear, which indicates that the Romans were making considerable advances on the original concept. But they did not use it. One reason is that the slaves employed in the grinding of corn would resist and, with a plentiful supply of slaves, no-one was particularly concerned to replace them with machinery. But a further reason is believed to have been the fear of interfering with nature.

John Reynolds, in *Windmills and Watermills*, puts it this way: "The new machine met with strong official opposition. It was claimed that the state would be endangered if this great labour force were to be thrown out of work." As to the gears, which must have represented an even more outrageous interference with the natural order, the Romans seem to have reconciled themselves to using them for some purposes where the end sanctified the means – purposes which, as Mr. Reynolds puts it, "seem strangely frivolous to the modern mind, such as the operation of temple doors or theatrical machinery. The ancients viewed the forces of nature with a deeply ingrained religious awe which tended to inhibit experiments with water power. When every river or waterfall was inhabited by its own *genius loci*, to set up a water wheel was tantamount to harnessing the gods for the menial service of man. But purely practical considerations, allied perhaps to a growing cynicism, ensured the gradual acceptance of the water mill." But mostly without gears.

Inevitably, the earliest acts of impiety took place far from Rome, in Britain. "There would have been some difficulty in supporting the large garrison on Hadrian's Wall with a relatively small local population to provide slave labour", says the compiler of *Industrial Archeology of Watermills and Waterpower* (Heinemann Educational Books/Schools Council). "Thus, there was an incentive to develop the latest labour-saving machinery before it was used in Rome itself and the sites of two mills have been tentatively identified near Hadrian's Wall."

When Emperor Constantine introduced Christianity, slavery was abolished and labour for grinding corn became short. In addition, the pagan worship of the water spirits ended. The way was clear for the development of water power.

It was the Romans who realised first the advantage of driving a mill by enabling the water to stream down on to the wheel from above, where it had the advantage of not only a flowing river or stream but also of the weight of falling water. The "overshot" water wheel, as we now call it, can produce efficiencies of 70–90%, while the original undershot wheel had an efficiency of only 30%. It is worth pausing to appreciate that it took probably 200 years before someone saw the point of this. In retrospect, it is self-evident. Yet even today most people think of water wheels in the way that most of them have survived, on the side of the old mill house with the water flowing placidly, and inefficiently, underneath.

This becomes significant today when many different devices are competing for attention in the race to capture wave energy. The schemes vary enormously and some of them seem so obvious that they naturally arouse scepticism. Why, one wonders, has it taken so long to build something so attractive, simple, obvious? But, as with the water wheel, the obvious is elusive. And we have not got 200 years to spare.

In Britain, the introduction of water power caused considerable social problems. The miller was a central figure in society but, in the Middle Ages, he had to hand over almost his entire produce to the Lord of the Manor who owned the mills, retaining only a small quantity as toll. No-one else was allowed to grind corn but, as always happens when the law tries to be all-pervasive, the peasants used quern stones secretly at home. On one occasion, the Abbot of St. Albans confiscated them from his tenants and used them to pave a courtyard. During the Peasants' Revolt, the stones were wrenched up and destroyed.

The wheel, so to speak, turned full circle when steam challenged water. For a long time the power of water was greater. As late as 1854, a water wheel was built by the Great Laxey Mining Company on the Isle of Man to pump water from a lead mine. It was 21 m in diameter and weighed 100 tonnes and produced 172 kW. Ten years later another wheel, 190 kW, was erected at Rishworth Mills, near Halifax. In the US, one water wheel developed a power of 7.5 MW. But steam was winning.

Its strength was that it could produce more power than water wheels in most countries. The drainage of land had changed the rivers, which ran at more varied paces and the domestic demand for water was increasing with the population. Looking back, one can see that industry had no real alternative. It would not have been possible to meet the demands of the 19th and 20th centuries without steam, however undesirable many of its consequences have proved.

Today, we are at another turning point. It can scarcely be doubted by anyone who has studied the facts that we have got to surmount or submit to a crisis caused by the dwindling of energy resources. There are many advocates of concentrating our major effort on nuclear energy and they will and do insist that its dangers have been over-stated. Ironically, the Wave Energy Steering Committee is based at Harwell, where it is sacrilege to suggest that radiation might be anything but good and cheap. And you are escorted to the office of the experts in charge of wave energy past locked cubicles labelled DANGER − RADIATION, and your guide will assure you that there is nothing to worry about, the place is the safest in the world. Hmm.

The supporters of coal are looking forward to the day when, having dug up the Vale of York to make the Selby coalmine − which is creating far more disturbance than the original inquiry suggested that it would − they can turn on the Vale of Belvoir. And the Coal Board geologists have been making some interesting discoveries in the Cotswolds, about which they are not talking much at the moment. Indeed, it is whispered that the city of Oxford itself is sitting on top of a rich seam and doubtless at the right moment we shall be assured that there will be no subsidence and all those old buildings are quite safe really.

The National Union of Mineworkers, as well as the Coal Board, seem anxious to condemn a growing number of men to spending their working lives underground, living dangerously and unpleasantly in a barbaric environment that no human being should ever have been condemned to.

Mr. Arthur Scargill, perhaps a little unexpectedly, has emerged as a champion of the environmentalists when nuclear energy appears to be a possible rival to jobs in the mining industry. It does require quite a mental effort to see an environmentalist who thinks of underground mining as a civilised pursuit. But there is no doubt that Mr. Scargill, despite his devotion to the interests of his members, would be prepared to see the last coal mine closed down and his miners employed in a less demanding occupation, if society could find alternative sources of energy and would ensure that the miners were properly re-deployed. He is a rare idealist among trade unionists.

Other potential opponents of almost any source of alternative energy are some of the dedicated environmentalists. They do not see it, or intend it, in that way, of course. Indeed, they champion the idea of seeking alternative sources. But many of them do tend to suggest that, even with extra sources, the end is near. They are the modern equivalent of the religious zealots in pagan Rome who believed that it was sinful to harness the god of water for the service of man. One may respect the motivation of the pessimists and admire their readiness to return to a pre-industrial condition. Unfortunately, there isn't room for us all to go back to nature.

There are now engineers and scientists who are working towards a solution that may prove as revolutionary as the invention of steam power. The Government, despite its caution, has agreed that for Britain wave energy is the most hopeful development. Against this small and unpowerful group are the modern equivalents of the slaves and peasants and mill workers who lost their jobs when watermills did the work and heedless societies ignored the victims.

The same could happen again, if society is unconcerned. Every technical development threatens jobs — or creates more leisure. Finding the right answer is a problem for all of us. Today's slaves and peasants will include nuclear scientists, leaders of the National Coal Board and the National Union of Mineworkers, the cautious instincts of civil servants and politicians and the muddled sentiments of people who distrust all industrialisation. They make a formidable combination. But the waves will not go back.

CHAPTER 4

How Many Gigawatts?

How serious is wave energy? Are we discussing a minor addition to the ways in which our needs are met by the various bodies who supply us — something like, say, a new telephone exchange which, we are assured, will help us to avoid calling wrong numbers and which, after a lot of upheaval, comes into being with no noticeable impact on the consumer? Or are we really discussing a major development in our technological history which can provide an answer to our demand for power when the oil gives out, and which can be of even greater importance to other countries whose resources of natural fuels are less than ours and whose needs are greater? The answer, which has come from the one source in this discussion that can be regarded as unbiased, is that we are talking about a system which could supply all our electricity. That statement requires a great deal of interpretation, insertion of ifs and buts. We need to subtract from it a number of factors; but we can also add to its conclusion and contemplate a future in which we might produce a surplus of energy.

This analysis is based on a study of wave energy by Ian Glendenning, Leader of Long Term Studies Projects at the Central Electricity Generating Board's Marchwood Engineering Laboratories in Southampton. The CEGB will have to answer a lot of questions if the lights go out and the conveyor belts stop bringing up coal from the mines. Mr. Glendenning and his colleagues have been studying natural energy resources and they are unprejudiced in the sense that they are not themselves promoting schemes and are not seeking Government money. A cynic might feel doubtful about claims from some of the bodies with a possible ulterior motive; but the CEGB has nothing to gain and indeed is heavily nuclear-biased. A report has circulated in its office suggesting that alternative sources are unlikely to contribute significantly to our supplies, but that we have to

28

explore the possibilities to demonstrate this to those people who are opposed to expansion of nuclear power.

Mr. Glendenning and his colleagues, who are not responsible for this report, have been studying the problem with the scepticism that comes naturally to scientists and with the caution that will always accompany the work of the men and women who have to provide the supplies at an acceptable price. A major contribution of the CEGB has been to insist continually on the need to consider the cost of any energy. The enthusiasts, equally understandably, have been more concerned to emphasise the need to provide energy at whatever the cost when the fossil fuels run out.

The CEGB came on to the scene early. Its first major report, by a team of eight scientists and engineers, concluded in 1975 that they had not found any new energy source "that appears economically attractive at the present time". They added cautiously: "Any of the resources that we have considered (solar, wind, waves, tides and geo-thermal) might under certain circumstances provide an alternative and there does not appear to be any one obvious choice." It was in April, 1976 that the Department of Energy announced that it had concluded that "energy in ocean waves is intrinsically the most attractive of these sources" and it was around the same time that Mr. Glendenning was coming to the same conclusion.

He is mercurial, volatile and forceful, exactly the sort of person who would be the first to say that the emperor had no clothes. He thinks quickly, challenges everyone he is talking to and then leapfrogs over your head before you can catch up with his previous argument. He is a fast, determined car driver and in this, as in his other expressions of exuberance, the style is the man exactly. He is capable of saying the unsayable and he has done so.

In an early public study of energy from the sea, published in *Chemistry and Industry* on July 16, 1977, he said: "If wave power can be economically harnessed, its development potential along the UK west coast exceeds the present installed capacity of the CEGB." And later in the same article: "There are currently an average of 120 gigawatts of wave power being dissipated on the beaches − nearly five times the average demand on the CEGB." Finally, in the most up-to-date account of development, presented to the Oceanology International Conference in Brighton in March, 1978, Mr. Glendenning repeated his claim to an audience including the most distinguished experts on the subject, referring this time to a "potential" of 120 gigawatts and adding: "We can at least be confident that a 20-30 km long system would represent a useful 1,500 megawatt power station." These statements represent a revolution.

Let us examine the details. A gigawatt (GW for short) is the sort of word that makes people's eyes glaze over. One gigawatt is 1,000 megawatts (MW) or 1,000,000 kilowatts (kW) — that is a million one-bar electric fires. And 120 GW is 120 million fires. The CEGB's present installed capacity is 60 GW and our average consumption is below 30 GW. Hence the calculation that the wave energy of 120 GW is nearly five times the annual average demand.

At this point, the layman is tempted to start wondering how we can export this new found wealth. Who, after all, wants five times our electricity? And it is here that the cautionary note must be entered. It has been put to me by several people, including Mr. Glendenning, and I do not wish to under-play their warnings.

Let us take, first the interpretation of Mr. Clive Grove-Palmer, the Government's secretary of the Wave Energy Steering Committee which is in overall charge of the whole project from its headquarters at Harwell. He puts it this way: "Divide your 120 GW by two for efficiency and by another two for transmission. Say your average energy is 50 kW/m, then from 1,000 km (600 miles) of devices, we would get only 12 GW." Only. Even on this rightly-cautious estimate, we are talking about replacing nearly half our normal consumption of electricity without using up fossil fuels. If that is not revolution, what is? Compare this with the extraordinary pessimism of Mr. Alex Eadie, junior Energy Minister, who has said that by the year 2,000, *all* renewable sources of energy might produce about 3% of the output of the national grid. Note, too, that there are experts who will argue that we would lose not 50% but only 10% at each stage, from generation and transmission.

Now let us take Mr. Glendenning's own cautionary note. He emphasises that his 120 GW refers to the whole UK west coast and much of it would be inaccessible for economic exploitation, as the line extends well north of the Hebrides and transmission becomes difficult, if not impossible. However, in his later paper, in March, 1978, he did contemplate extending the operation by placing devices off the north-east coast of England, where it would be closer to industrialised areas needing power.

He argues that 50 kW/m is the maximum continuous rating of a smoothed system in which peak input is at 150-200 kW/m, "because of the random nature of the sea". The load factor would be 60-70%, so the average output would come down to 30-35 kW/m. With conversion at 80% and transmission at 90%, then the average *landed* power would be 24-30 kW/m. So, as he notes, one would come down on his estimate to much the same figure as Mr. Grove-Palmer — that is, 25% of input. He adds that this could be misleading because one might obtain as much as 60% under desirable conditions and only 1% or 2% in extreme conditions when the devices could not absorb the waves because they were too powerful, or when the waves were too placid. He concludes that from 1,500 km of device, one might achieve 36-45 GW. This is not far different from the slightly more cautious estimate of Mr. Grove-Palmer, based on 1,000 km, when you take into account Mr. Glendenning's insistence on a peak input of 150-200 kW/m. He too comes down to roughly 25% of the power received by the wave energy generators.

Mr. Glendenning has also made to me the point that as data on inshore wave power becomes available, it is becoming clear that the energy available is lower than at Station India and that this may mean that devices will be smaller; this could reduce or even improve the cost-effectiveness of the system. And he adds: "One thing I have learned in the past four years is that wave power is full of surprises and there has been much eating of hats already and there is more to come."

I have given at some length the cautionary tales because it is not my purpose to pretend that a simple solution to the energy crisis is just waiting for some Minister to wake up; if only it were as simple as that! No, there are problems which are being tackled — too slowly, in my opinion, without the sense of urgency demanded by a situation in which we are, as Cockerell has put it, living on our capital.

To keep the balance, let me mention one aspect of the problem which seems remote now but is emerging as more and more significant. What is the objection to a line of devices behind the "front" line at sea? Supposing we built a line of full-scale generators out in the Atlantic, the sea behind them would be calm for some distance. It is generally accepted that a new fetch would start on the line of the generators. It would then take 160 km before the wind had whipped up the sea sufficiently to make it worth-while putting down a new line of devices. But the idea that is

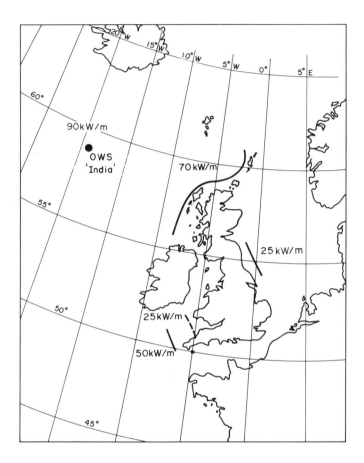

Figure 4.1   The map shows that the power available varies from 90
kW/m out in the North Atlantic, at Ocean Weather Ship *India*, to 25
kW/m off the north-east coast of England and in the Irish sea. The
accepted estimate is that this is a potential of 120 GW, five times the
average demand on the CEGB.

gaining support is that it should be possible to put down a line of half-scale generators in the calmer sea only 80 km behind the first line. As Mr. Glendenning hints, it is too early to know just what the cost-effectiveness would be. The plant would cost less and would produce less power. But the problems of construction and mooring would be simpler. So that one could then think in terms of an expensive first line producing something between 12 GW and 45 GW, and a second line, easier to service and cheaper to build, producing comparatively less. Then it would be open to the CEGB to calculate which should be considered as our best plants for high load factors and which for peaking duties — to put it in conventional terms.

There is also the growing feeling that we might be best served by devices smaller than those now regarded as full-scale, based not off the Outer Hebrides but rather in calmer waters where they would produce less power but cost less to build and service.

Those are the two sides of the coin — the prospect of less power than we may hope for; the alternative of a growing series of power stations of different sizes in different conditions for different functions, combining with the major producers to give a flexible source.

This, I hope, does give a fair interpretation of the options available and does put in perspective the figure of 120 GW which, set by itself, would be misleading. But we should bear in mind that although it might have to be cut down to 25% or even lower, it could equally prove an under-estimate once our wave energy programme begins to move.

I have devoted a lot of space to placing the figure in what is, I hope, a right perspective; we should now return to Mr. Glendenning's paper which is the most important document in the present whirlpool of discussion, not only because of the source from which it comes but also because it is the most recent, authoritative survey of the subject. It comes at a time when it is fashionable to denigrate the prospects of unconventional sources. The official attitude of the Department of Energy, as stated by Dr. Freddie Clarke, Research Director (Energy) at Harwell, and the key figure in the Government's thinking on our future supplies, is that alternative sources are "likely to make only a minor contribution" this century but that there could be "a take-off in their use during the first decades (plural) of the next century". If one accepts this, then it is difficult to contest the conclusion of the Department, and of some people in the CEGB, that we must extend our dependence on nuclear energy. This is taken as standard wisdom by most of our scientists and electricians, who will tell listeners repeatedly about the convenience of being able to press a button and bring in cheap power.

Yet in recent months I have been able to put it to some of our leading experts on energy that the waves could provide all the electricity that we need, plus petrol to run our motor cars in a non-polluting "hydrogen economy" (about which, more later). The reaction has ranged from a scathing glare to a menacing demand that I should "say that again" from a nuclear scientist in the Department of Energy. In fact, no-one is urging that we should abandon all sources of energy other than the waves. What can be asserted is that there is enough energy available to enable us to regard it as a prime source of power. Let me now summarise the main points of the Glendenning report.

Mr. Glendenning notes that the first practical development in Britain was the demonstration by Stephen Salter of a duck with an absorption efficiency of more than 80%. "Both the scale of the available power and the modular nature meant that here was a potentially valuable source of energy which could be researched and even tested at full scale without risking the very large sum of money which, for example, a meaningful tidal power programme would require." By reference to "the modular nature", Mr. Glendenning is referring to an aspect of wave energy of great significance for the shipbuilding industry. It means that once a decision has been made about a device, it will be possible to reproduce it cheaply. Unlike a tidal barrage, which is a once-off operation, wave generators can be repeated indefinitely and we could start off with, say, one raft or one string of ducks and then expand production to whatever extent was justified by results.

Mr. Glendenning paid considerable attention to what is, for the CEGB, a key factor, the cost. There are others who will argue that this is secondary when we are discussing not how much money we are ready to spend on electricity, but on whether we have any at all. It is necessary for politicians and the nationalised industries to satisfy themselves on this aspect. As Mr. Glendenning puts it:

"The preferred plant will always be that which will enable the generating system to continue to meet a changing demand at the minimum overall cost. Each new plant is therefore assessed for the balance between the cost of installation, operation and maintenance and the overall savings, primarily of expensive fuels, which it would be expected to make in normal operation over its life.

The pattern of electricity demand is such that these assessments lead to the development of an optimum mix of generating plant in which high capital cost/low fuel cost nuclear plant is employed for high load factor duties, intermediate capital and fuel cost plant is installed for mid-merit operation and cheap quick-build plant such as gas turbines are installed for peaking duties despite their high fuel costs.

Although the fuel cost will be zero, wave power stations will be expensive to install and maintain and must, like other high capital cost plant such as nuclear power stations, be operated at a high load factor if they are to be able to recover the investment through fuel cost savings."

I quote this at length because it is an excellent exposition of a view for which I have no sympathy and which, in fairness, must be presented fully. To put Mr. Glendenning's argument crudely it is that a nuclear power station costs a lot to build but, as the juice flows cheaply, it justifies itself so long as you use it a lot of the time. By contrast, gas turbines are cheap to build but the fuel is expensive so it makes sense to turn them on only when there is a peak in demand. Given that wave energy generators will cost a lot in capital, but nothing in fuel, they make financial sense only if you can keep them running almost non-stop, as competitors with cheap nuclear energy. My objection to this is that it ignores the question of "cost" in social terms. We know that fossil fuels, and uranium are finite and will sooner or later give out. We know that they are polluting and dangerous. There is no mathematical equation to take account of these factors. But if money matters, so do the other considerations.

Then there is his second point that wave energy will not be available as readily as other forms. "Most power plants", he argues, "are available to produce power on demand, a net availability of close to 100%. Wave power, however, like wind power, will be subject to the vagaries of the weather and therefore could not achieve such a high net availability unless it were to be rated at a very low power level indeed." My first response to this was to mention that nuclear power stations were liable to be out of service, when something went wrong, for 18 months at a time and only between 40% and 60% are working at any one time. Mr. Glendenning rounded on me with his eyes gleaming. "Ha", he said, "you've been talking to Stephen." I had indeed seen Salter. The wave energy family is incestuous. Mr. Glendenning's reply was that some nuclear stations were less reliable than others but he remains emphatic that nuclear power is the most convenient source. Indeed, his point about the variations in output from the waves has become more significant to everyone after the initial euphoria. It was at first said, rightly, that the waves were at their most ferocious and most productive in winter, when we most needed energy. But with the detailed studies that have taken place of the spread of supply, it is clear that even in the depths of winter it is possible to have little energy available from a sea as turbulent as the North Atlantic. One can have two or three days when the sea becomes unexpectedly calm. The answer to this problem may be to spread out the devices, as it will be rare for the North Atlantic, the North Sea, the Atlantic off the Scilly Islands, the Irish Sea and the approaches to the Bristol Channel all to be calm at the same time. Mr. Glendenning has himself helped to answer his own point. In July, 1977, in his article in *Chemistry and Industry*, he produced a map showing wave energy sites off the Outer Hebrides and the Scillies. Nine months later, he showed new sites off the north-east coast of England and across the Bristol Channel and the Irish Sea.

A further answer is to extend our use of pumped storage, which will be dealt with later. For the moment, it is worth noting that we are at present feeding our limited storage facilities by actually wasting energy. It needs four units of electricity to pump water uphill in order to produce three units as it comes down. With fossil fuels, this is a wicked waste; with the waves, the input is free – "of no concern when the gods pay for the waves", to quote Salter.

Yet we must still grapple with the problem of what to do on a day when the sea is calm in the North Atlantic and not particularly fruitful in other areas, and working wives return home around 5 p.m. and turn on their electric fires and kettles and ovens – or, more simply, when a popular television programme ends. How long, then, would pumped storage prove adequate? At present, it would last for about five hours and next day the country would be in trouble. It is then that one would be grateful that we had not shut down all the other sources of power, benign and malignant. We should then indeed call on the conventional and perhaps nuclear sources. No-one is suggesting that we should abandon all but the waves as our power supplier. Where public opinion needs to challenge CEGB wisdom is in the area of cost. If we have to pay more for wave-electricity, then we must recognise the fact; the alternative is to burn up the possibly cheaper, limited fossils knowing that the joyride will one day lurch to a stop.

To resume with Mr. Glendenning's report. He is emphatic in insisting on devices that will cause the smallest loss of output for repairs and maintenance. He calculates that "a loss of two months' output per year through a combination of repair time and waiting for 'weather windows' and operating and maintenance charges of only 6% of the capital cost per annum, could double the cost of the energy generated".

Mr. Glendenning presents his arguments with vigour and eloquence and corrects himself apologetically if he feels that he has over-stated his objections to the work of colleagues whom he respects. His enthusiasm, the sharpness with which he corrects it, the application of scientific method to what is, at this stage, a romantic notion make him an interesting example of the stirring emotions that eddy backwards and forwards among all the people close to the project.

He and his colleagues at Marchwood have made one valuable contribution of their own. Early experiments by Sir Christopher Cockerell were with a line of seven pontoons in one unit. The CEGB has been able to demonstrate that small may well be energetic as well as beautiful. It suggested reducing the number of pontoons to three or even two, and holding one of them stable to produce the contrast in motion and thereby the power. The increase in efficiency has been described as dramatic. The size of the Raft could then be reduced from 120 m to 80, with considerable savings in cost of production. But Mr. Glendenning admits that this would provide an overall efficiency only "approaching that" of Salter's Duck; the general picture is that the Duck would be probably more expensive to build and maintain but might well be the most fruitful device while the Raft would be cheaper and easier to build and service, while producing less electricity. Ideally, one would like to see Ducks in the North Atlantic and rafts in the North Sea.

Mr. Glendenning contributes an illustration of the caution that has governed his own approach. He notes that the Ducks would be on a spine of 15 m diameter "which is very large by normal engineering standards and not well suited for sea water sealing or bearings of any standard type. Without seals between Duck and spine all equipment will be forced to operate in sea water leading to very uncertain component life." Well, yes.

But note the difference between his remark that the size of the spine means that it is not "well suited" to sealing, and his leap forward into saying that "without seals" the troubles would be considerable.

I have left to the end his clearest statement: "a 20–30 km long system would represent a useful 1,500 MW power station". Taking his least favourable figure of 30 km, that is to say the longest length, it would mean that our average consumption of 30 GW would require 600 km of wave energy devices, to produce all our normal needs – less than the 1,000 km normally named by the most optimistic inventors and their supporters. By now, readers will appreciate that the figures need interpreting with scrupulous care. Mr. Glendenning puts the mileage of devices as more than twice this figure to distinguish between the rating of a station and its average output. One returns to something near his calculation of 1,500 km. He adds that there would still be the problem of guaranteeing supplies when the waves were calm. These points have been covered in this chapter and will be considered again later, when we discuss pumped storage. With all his provisos inserted, we are still in possession of a statement that gives a mental image of the type of power plant that the waves can provide.

His major conclusions are the essence of the case for wave energy and should be written up large on the walls of the Department of Energy:

> (1) "All the research to date suggests that wave power will be technically feasible and could conceivably be economic.
> (2) Considering the UK coast alone, there is a 'potential' of 120 GW, five times the average demand on the CEGB system.
> (3) A 20–30 km long system will produce smooth power and, at 50 kW/m would represent a useful 1,500 MW power station."

Discuss these statements as you will, reduce them for losses in capture of energy, generation and transmission, and still you are left with the hard fact that the latest and best-informed assessment to come out of the CEGB makes it plain that the waves can produce all the electricity that we need.

CHAPTER 5

The Letter from the Bionics Department

The letter which Stephen Salter sent to Whitehall eventually arrived on the desk of Mr. Gordon Goodwin, Principal Scientific Officer at the Department of Energy. One's first impression of him is of a man withdrawn and taciturn and the reason for this attitude, presumably acquired in the job rather than inherited, is obvious:

> "We get a lot of oddball inquiries and schemes into this department, and most of them come my way. We frequently get perpetual motion schemes. Lots of people want to do something about using sawdust. Or painting the moon another colour. And during the oil crisis, there were some very strange suggestions. Salter's letter was from a peculiar address: the School of Artificial Intelligence, Bionics Research Laboratory, Edinburgh University. The letter had been photostatted at some stage on its rounds and the thought crossed my mind that if someone was trying to do a hoax it would be easy to cut up a potato and 'print' a heading and send off a a letter like this.
>
> So first I phoned the university and asked if there was someone there called Salter. The switchboard said that there was. I rang off and thought. Then I rang back and spoke to him direct and asked some questions. I asked him did he know what the wave data was like and did he know the difference between winter and summer loading and the best area for wave energy and he came up with the right answers. He *had* covered the ground."

So, it should be mentioned, had Mr. Goodwin. He had investigated wave energy more than 12 months before Salter wrote. It is not widely known that the Government awakened to the need for alternative sources of energy a year before the Middle East war had brought home the seriousness of the situation. To give credit where it is due, in early 1973 Lord Rothschild and the Central Policy Review Staff, as the Think Tank is known officially, were asked to study the question and Mr. Goodwin had been the man who had looked at wave energy. He had only seven weeks to draw up a paper while carrying on with his normal work. He estimates that he had three man-weeks to investigate what was an idea 200 years old and also, in terms of modern technology, brand new.

His first discovery was how little was known about the subject. "People's ideas of a wave were completely wrong. For a start, it is a very flat thing, drawn to scale, hardly a perturbation." That is, the Atlantic reaches down to 10,000 m; a wave will rarely reach 30 m. A pimple.

Mr. Goodwin made the rounds that, since then, have been followed by most people concerned with wave energy — the Institute of Oceanographic Sciences, the Hydraulics Research Station and the research associations dealing with the Merchant Navy, the Royal Navy and naval construction.

> "From my work on North Sea gas, with undersea piping, and being not long out of commerical practice, I devised a scheme which could have been built off the shelf. It was based on floats. It was just a constrained float driving a pump into something like a Pelton wheel.
> To be confident that the system was not going to be subject to serious criticism, I taught myself a fair bit about the waves. I got a feel of them. I learned things from them. I learned to do a technical assessment of wave energy. The first thing was to do calculations on how much energy was there. It was such an enormous figure that I thought I must have got it wrong by a factor of 10 or something. I tried other ways. And I became certain that there was a significant natural resource. The report was produced and I went on to other things. Then came the Middle East war and the letter from Salter."

What Mr. Goodwin did not know at that time was that Salter was imaginative enough to realise how his letter must have been received. He has recorded the facts.

> "About 10% of the population of this country suffer, at one time or another, from some form of mental disturbance. In a fair number of them, the manifestations involve writing to Cabinet Ministers. In normal times, the mail is evenly distributed among the various branches of government. But at the end of 1973, the trousers-for-dogs and bring-back-the-cat letters faded away and with one mind all the writers switched to helpful proposals for solving the energy crisis: treadmills for fat business men, sewage-driven cars .... I chipped in with a short letter to Peter Walker who was then responsible for energy. In due course, I received a polite acknowledgment. The fact that my note had come from the School of Artificial Intelligence and some quirk of my literary style suggested that the whole business was a cruel hoax against Peter Walker when he was battling with Joe Gormley and the Arabs. Their initial approach was very tentative. But they were soon in Edinburgh ...."

"They" refers to Mr. Goodwin. We met in his large, agreeable office in Millbank. It is elegant and comfortable apart from one detail. An old tin wastepaper bin keeps the door jammed open on to the corridor. He apologised, explaining that he suffered from claustrophobia. This is understandable. The combination of roles as backroom scientist, expert on oddball inquiries and civil servant is justification for any psychological quirk. He is a chemical engineer and asserts that this is the ideal background because his training embraces every form of engineering and enables him to talk to mechanical, constructional and electrical engineers on their own subjects. He also has a degree in biology, has worked on computers and spent 12 years in industry, in Britain, Europe and New York before entering Government service. "I came to the Ministry of Power (as it was then called) on the basis of my industrial/economic experience", he recalled. "Yet the first thing they did was to send me off on a course in economics." Someone, somewhere was determined not to have any new ideas upsetting the established order.

After satisfying himself that Mr. Salter knew something about the waves, Mr. Goodwin asked him: "Have you got a device working in a tank that is absorbing wave energy. I thought he would say No. But he answered Yes, I have. I said I'd better come up and see him." Mr. Goodwin was on the first available train to Edinburgh. He learned how Salter had tried "various shapes" before arriving at the now-familiar Duck's beak with the rounded bottom which absorbs energy without displacing water and creating new, wasteful waves of its own. And he saw it at work in a tank. Mr. Goodwin recalled:

> "The full importance of what I saw did not sink in until 24 hours later. I was back in London and I realised that I had seen an artificial wave arriving at the beak and calm water on the other side of it. You did not get a strong reflected wave going backwards, as you would with a breakwater. This meant something close to total absorption of the energy. He could measure what was going in and what energy was being absorbed and the result was 90% whereas previously one had thought about capturing perhaps 30% of wave energy. And Salter's figures were borne out by the still water on the leeward side of his device. So here was confirmation of what the instruments showed. You were getting a powered take off. It was the first time that a measured high efficiency wave generator had ever been exhibited. It was a startling thing to see and I was the only person who had seen it, apart from Salter and his colleagues.
>
> Anything like this that could be made to work had to be taken seriously. It completely altered my view of the concept. But at this stage, there was hardly another soul anywhere that I could turn to. Everyone laughed at you – except Don Gore (who, as Deputy Chief Scientific Officer, was Mr. Goodwin's senior). He always felt that wave energy did carry credibility. He felt that it was ridiculous that with all our modern technology we could not do something with this. When I was worried that I was getting the energy levels too high, it never came home to him that such a thing was possible.
>
> So, after thinking about it, I rang Salter again. I told him to take a

photograph of his device. He said he hadn't got a camera. I said, get one, borrow it from someone. He said he didn't know anyone with a camera. I said I'd seen someone up at the university carrying a camera when I was there. He said he didn't know who it was. And anyway he hadn't any money to buy films. I said steal them or borrow money but I must have some photographs. And then he did it. And it was the first time that anything like this had been seen in the Department of Energy.

Then I said I wanted a cine film. He got me one. We were lucky. We had in the Department a projector which you can stop and play it backwards. And when we did this with the wave hitting the duck we saw some quite astonishing things. It showed us just what was happening, how the wave energy was being absorbed.

The way Salter had arrived at his success was by doing a very clever thing. Before, people had worked either on paper *or* in the water. Salter worked on paper and then dipped his paper in water. He went between paper and experiment. He had a tank of water in his laboratory. He looked at how a device worked, tried it experimentally and finally arrived at this profile.

There were a number of institutions aware that we had a crisis but not a dicky bird had come from them on wave energy. A complete outsider like Salter had gone round to Woolworth's, bought sixpen'orth of Balsa and made fools of all of them. This is a way of influencing people but not of making friends" – an oblique reference to Salter's critics in such places as the Central Electricity Generating Board and the National Engineering Laboratory.

"All this while", Mr. Goodwin continued, "Salter was applying for a grant from the Science Research Council, because he was an academic and at university and this was the natural place for him to apply. Very unfortunately, he had only recently been working on a project which had been aimed at producing an artificial engine. In the School of Artificial Intelligence they had built a robot which could assemble a toy steam engine from its component parts. You could put the parts in any configuration and you would end with this toy assembled. Salter had, with his own hands, built the robot part of it. This was the successful part. But there was a hitch. On the instruction, Put all the pieces away, the computer dismantled the engine and put the parts in a box and then went on trying to put the box inside the box! There had just been some laboratory computer software needed to make the experiment a success. But by then the Science Research Council had spent about £2.5 million on it and they just decided to wind the whole thing up.

So as soon as you said to the SRC that Salter had been working in the School of Artificial Intelligence, down the shutters went because of this sad experience."

Mr. Goodwin managed to say this without bursting into laughter and I kept a straight-ish face, too.

Mr. Goodwin, however, took up the battle for Salter with the SRC, saying that the Department of Energy would like to see him funded. He eventually obtained an offer for Salter, for "a few thousand pounds", which Salter never took up. One understands his feelings.

He had made a staunch ally in Gordon Goodwin, who then tackled another Governmental body, the Mechanical Engineering and Machine Tools Requirements Board of the Department of Trade and Industry which can claim to be the body which has made wave energy possible. Requirements Boards are a most interesting branch of Government. They were introduced in 1972 by the Heath Government with the purpose of "identifying requirements" in research and development for both manufacturing industry and research establishments where the Boards consider that there is a likelihood of technical success. They take into account numerous factors, among them reduction of pollution and energy conservation. There are eight of these Boards covering most sectors of industry and they match the thinking of Mr. Wedgwood Benn in one important respect: if they give help to a project which becomes profitable, they can ask for a share of the winnings. Yet it was a Conservative Government which introduced the scheme. "You could regard its main function as priming the pump", said a spokeswoman of the Department of Trade and Industry; it is interesting how many of our well-used phrases come from water power.

The Boards are allowed to award up to £50,000 a year for any project at their own discretion. They gave Salter £64,000 but, by spreading it over two years, were able to pay the money without asking for a Minister's approval. If you recall how difficult it was at that time to persuade anyone that wave energy was a serious proposition, you will understand just how smoothly the civil servants managed to ease the project through.

Meanwhile, the National Engineering Laboratory at East Kilbride was producing a long survey of wave energy. It is probably still the most comprehensive work on the subject. There are about 50,000 words of text, numerous graphs, charts and illustrations and an extensive list of sources. Most people have regarded it as the standard work.

Mr. Goodwin was less enthusiastic. "We granted them £13,000 to produce this and they had access to Salter's work. It is horrible. It is rubbish. It is paper ideas. I had argued that water would not do what you expected. I thought that we should get our feet wet. I was over-ruled and work went on at the NEL."

The outburst seemed at first surprising. The NEL enjoys a high reputation and personally I have always found its engineers helpful and, insofar as one can assess the competence of a group of brilliant innovators, of the highest technical quality. Most engineering students, particularly in Scotland where engineering abilities are so frequently found at their best, regard an appointment to work at NEL as their highest ambition. Why then was Mr. Goodwin so critical? It is obviously in part an expression of the competitiveness and tension that are rising throughout the wave energy world as we get closer to decisions that will be critical to the nation's future. But there is another aspect to it: the NEL started out as advocates of Salter's Duck and then moved towards favouring the Oscillating Water Column which was at that time known as the Masuda concept. It settled on this scheme "for further technical and economic assessment". And today the great resources of the NEL are behind the OWC, while Salter has to rely on considerable but less well-equipped resources.

The NEL will not be surprised to discover that in the sharpening race for a breakthrough in a new technology, passions are rising. I have been told by its engineers there of some of the opposition that they have encountered when, for instance, they have been called in by manufacturers to help in de-bugging new machinery which refuses to bed down. No-one likes the outsider or believes that he might be able to do something that had proved insoluble to the residents. Yet the NEL, by hooking up machinery to its computer, has often been able to analyse a problem and suggest a solution that had eluded those experts who were most familiar with the particular machinery.

Its reasons for favouring the OWC will be examined in detail later. It is certainly one of the most promising schemes, though it has its drawbacks: for instance, the Salter supporters will emphasise that the Column has to withstand a far heavier battering from the sea than the Ducks which flip up and down in the waves. Against this, the Column has the advantage that it is already working. Masuda, the Japanese inventor, has produced navigation buoys powered entirely by wave energy. They are functioning at sea and are shortly to be manufactured in Britain. And Mr. Goodwin himself mentioned that the discovery that these buoys were working, and working well, "at a point where credibility mattered", was one of the events that made him interested in wave energy.

The incident helps to illustrate what in industry is called the "creative tension" which is impelling forward the whole project. It remains "creative" so long as the project moves forward; it is when backbiting and inward pressures take over that a neurotic atmosphere develops and that is when a new concept is in danger. It has not happened to wave energy yet but it will do if the Government's financial stimulus falters.

Returning to Mr. Goodwin, he has determined ideas on the future of the project. He has precise ideas about the speed at which it should go and the limitations which should be imposed – or accepted. "It is a very extensive resource in national energy terms" he said. "It is modular and so has an advantage over tidal barrages. You can stop. Or extend it. Or build it up after 10 or 20 years. It peaks with the winter demand and shuts itself off in the summer." That is the positive side.

And then, prompted by my argument that it could become our basic source of energy, he gave the other point of view.

He does not accept, as the CEGB is now starting to do, that by stationing your generators off different coasts, you could be certain that there would almost always be wave energy pumping ashore. "It must be off the Hebrides" he insisted. "Nowhere else is significant." I put it to him that even if one achieved a smaller number of kilowatts per metre of device, it would make sense to use the energy of, say, the North Sea or the calmer waters off the Scilly Islands as alternatives. He disagreed.

Why should it not be a basic source of energy? "You cannot make an unfirm source into a firm source" he insisted. "Forget it." I have since spoken to other engineers, including Salter and a very senior figure inside the CEGB, who contest this argument. With pumped storage, the sea can be made into a "firm" source; indeed, some countries have already done something similar with hydro-electric power.

Why not build a full-scale prototype now? Here he was particularly insistent. "We are talking about what is probably the most awful engineering job that anyone has ever taken on. There is no device that has not got real and intense problems. If and when all systems are fully engineered, you will find that each has its strengths and weaknesses. It is far, far too early at the moment to start saying what is the obvious line of approach. I am fairly certain that if and when wave energy devices get deployed, the first generation will be expensive, unreliable and inefficient. And if they are not, by some miracle, they will be different from every other engineering development we are used to."

Another scientist, one close to the centre of the wave energy programme, and who cannot be named in this context because of his role in advising on future policy, is less sweeping in his criticism. He is an advocate of the Oscillating Water Column which, as the Japanese have shown, is a practical proposition. It could prove expensive in construction and less fruitful than others in its productivity but it should be cheaper in maintenance and less demanding in any novel technology – everything about it is standard engineering. My friend puts the Cockerell Raft as a close second and the other two far behind. The Hydraulics Research Station Rectifier demands an enormously strong building, standing on the seabed, with non-return valves the size of a shopfront window. My friend told me, with animation about his objections to the Duck. "Salter does not deserve even a tick", he said, as he chalked away furiously on a blackboard with coloured diagrams to explain his views. "The Duck is complicated to build while the Raft could be constructed on any slipway in the country. How do you manufacture the gears, the power take-off and the drive that Salter would need? Imagine the problems when a Duck goes wrong. The repair work would be below water on a device as big as that" – pointing at a large laboratory. "You have a mechanism moving up and down, mostly below sea level, and you would need divers to work on it beneath the surface while it rolled."

I put it to him that the more ardent supporters of the Duck argued that there would be little needing maintenance. He retorted: "Show me a ship that goes to sea without someone going around all the time with an oil can." Well then, it has been argued, notably by Dr. Norman Bellamy, head of the Electrical and Electronics Department at Lanchester Polytechnic, that the "float" or nose section of the Duck, which does the bobbing up and down, can be unclipped from the spine in ten minutes and towed away. My friend swooped.

We were lunching in a large restaurant and he leaped out of his chair and marched from one end of the room to the other, counting loudly as he did so. "One, two, three . . . fourteen steps." Then: "You see, this restaurant is 14 metres from one end to the other. The diameter of the spine alone holding the Duck is 15 metres. Then you have the outer structure. Do you realise now the size of the operation that we are talking about?" I got his point, though some of the other diners appeared puzzled.

In defence of the Duck it must be said that the part which has to stay fairly stable, the spine, is smaller than the bobbing cone. In this it is different from all the others which need stability for their biggest components: the outer casing of the Oscillating Water Column, the whole "block of flats" of the Rectifier and the rearmost pontoon of the Raft. So the problem of holding them steady will be that much greater than for the spine.

But to return to Mr. Goodwin (who was *not* the friend that I referred to in the preceeding section), he is the best ally of wave energy inside the civil service but he did make one statement that worried me. He is a gifted engineer who deserves most of the credit for winkling the first money out of a Government body to help Salter. But he said to me: "It is the judgment of the people concerned that the project is expanding as fast as prudent. *We would be negligent if we did not establish whether it was possible.*"

Here, I fear, is an unfortunate echo of some of the thinking behind Government policy, as inspired by the civil service. The Government dared not turn its back on wave energy when the project was presented with insistence by Salter, Goodwin, Cockerell and the rest, and when the energy crisis was ringing alarm bells in quiet corridors where chaps prefer not to make a fuss. The Government did not wish to be accused of indifference to our futures. Nor did it wish to arouse public anxiety and opposition with a nuclear-based energy policy. So it assigned what seemed to be a reasonable sum of money in order, it may have hoped, to deflect criticisms of negligence or indifference.

It launched itself, with a tolerable show of goodwill, into programmes designed to save fuel: double-glazing and insulation and solar panels and windmills on top of every cowshed. But it shied away from the really big decision, the development on a major scale of the energy that surrounds us, the waves which pound against our beaches and cliffs and are available for the taking – provided that we are prepared to invest the money needed to capture and transmit their power.

I do not believe that the Government's attitude is adequate for the next stage of development of a new source of power and in this I am encouraged by a scientist of eminence who is aware of the political significance of the issue. He is yet a third character in this chapter and I apologise to the reader for omitting his name, too. He is engaged in energy research and I have had to choose between using only formal sources of information and going behind the scenes to obtain unofficial guidance, with an assurance of confidentiality. I have chosen the second course because it yields the information that we need.

My second anonymous source is convinced that two strands of thinking are intertwining in order to encourage a cautious creep forward. "The people experimenting with devices", he said, "were very insistent that the Government should help when they were getting little or no money. But now they have a comfortable arrangement. And the last thing that some of them want is for the Government to move in, in a really big way, with perhaps £1,000 million and say 'Right, let's get some of these things into the sea and find out how they function.' That would mean Government control because the Government is not going to hand over that sort of money to private consortiums. And then there are the civil servants who are naturally reluctant to take a big jump ahead. But what we need is to get something into the sea on a full-scale project and find out just what it can do."

My source is one of those people who believe that the need to provide productive employment should be a major consideration in our future policy on wave energy. I put this to Mr. Goodwin. "Ah", he said, "now you are talking about a Public Works Programme. That is not really my subject".

Agreed. But it is ours, along with our future energy.

CHAPTER 6

The First Big Scheme

The pioneer, this century, of wave energy on the big scale at which it is being contemplated now is a civil/electrical engineer who learned about the power of water while working on hydro-electric schemes in Scotland in the 1930s. There have been several small schemes for the use of wave energy — one in Monte Carlo, and the Japanese navigation buoys. But Walton Bott came within reach of launching a scheme which could have produced all the power needed for the 800,000 people who live on an Indian Ocean island, Mauritius, of only 2,000 km^2, making it one of the most densely-populated areas in the world, roughly twice the density of the UK. It has no fuel of its own and could have led the world in producing benign power from the sea when, in 1966, the decision was taken to abandon work on the Bott plan because, strange as it now seems, the world price of oil was falling and it was cheaper to bring the oil almost 5,000 km from the Persian Gulf than to build a wave energy plant on the beach.

The Bott plan was based on a natural advantage enjoyed by Mauritius. The island is not only surrounded by water; it has a fringing reef which offers, literally, a stepping stone, saving the cost of under-water foundation work.

Mr. Bott went to Mauritius in 1953 to set up an electricity board after having spent many years in Scotland working on hydro-electric schemes which are still producing cheap power. He set out to plan the island's future energy resources and was given financial support by the Crown Agents, who have been getting some bad publicity recently; their better work is scarcely noticed at home.

Mauritius has a small tidal range and its problem of capturing sea power is different from Britain's but might well prove significant to many countries which could build their devices on or close to the beach. Mr. Bott set out methodically, reading all the literature available on waves and tides, which goes back to 1848 when Sir George Airy pioneered the study in his book *Theory of Water Waves*. Information on wave *energy* was effectively non-existent. Mr Bott also approached the Hydraulics Research Station at Wallingford, Oxfordshire, which is today working on a scheme that is a development of his ideas.

He decided that, at least in the Indian Ocean, the waves that must be used were those breaking on the beach because there was no way of building structures that could survive the force of the open sea. He lives now in Winchester, in semi-retirement as an adviser to the Crown Agents and he also lectures and writes on wave and other forms of energy. In his study, surrounded by books on water power, he told me: "I came to the conclusion that you could not put anything mechanical or electrical in the open sea. In Mauritius, we experienced cyclones which could bend 25 cm girders double. In Scotland, a 7,000-ton breakwater was rolled over. It was replaced by a 14,000-ton one, which was also destroyed. An American heavy cruiser lost 30 m of bow from wave action."

He is also sceptical of the prospect of capturing energy in deep water off Britain. "You could have two 30 m waves meeting in phase. That is a force that could move St. Paul's Cathedral. The only way to protect your structure would be to submerge it in time, if that is indeed possible. You cannot compare it with an oil platform which is solidly based on four legs and which is designed to let the waves through. It is not there to collect and process them.

"Then you have the problem of transmitting the power to land by cables not only resistant to sea water and sea animals but also presumably capable of dealing with violently-moving cable connections at the sea end." He is not seeking to belittle alternative schemes which come from people with a different engineering background. They believe that the problems he and others have listed can be overcome.

The Raft is an obvious candidate for survival – Kon Tiki demonstrated that. The Duck has the advantage that its biggest part, the beak, is designed to flip up and down in the waves. It would also be part of a line of perhaps 50 Ducks so that if one were damaged it could be taken out of the generating system: the electrical connections are being arranged "like the lights on a Xmas tree", as one expert has put it, so that one can go out without shorting the whole system. It is also thought possible that a Duck could do a complete somersault and still function in the waves coming from the opposite direction. And there are also systems being devised which would enable a beak to be unclipped from its bearings and towed away as soon as there was a "weather window". The OWC may present more problems because it is estimated that one unit – a Masuda "ship" – would need to be about 220 m long and 33 m in beam to be comparable with one Raft. This will be a problem for the National Engineering Laboratory, who are perhaps the world's leading experts on mooring; they are not sanguine about the size of the problem.

As to the cable, this is less of a problem today than it was at the time of Mr. Bott's early experiments. Pirelli have devised a method of manufacturing cables in lengths of 80 km – an important development because joints are always a weak point. They have experience of laying submarine cables going back many years. As long ago as 1965, they laid a cable of 119 km from the Italian mainland at Piombino to Corsica and Sardinia, able to carry 300 MW.

By coincidence, they have recently been working on a cable that will answer many of the new problems and it is, ironically, nuclear energy which has led them to it. An American company, Power Service Electric and Gas Co., asked them to produce 345 kV a.c. submarine cables from floating offshore nuclear power stations. The distance would be only 7 km and the cables would be embedded in the sea bed.

There is no problem about manufacturing it for that distance without joints but the final cable section to the floating power station provides exactly the problems that will have to be solved if we choose floating wave energy generators. The cables must be able to follow the platform as it moves in the waves and tides. The solution chosen by Pirelli is a flexible cable with a corrugated aluminium sheath, instead of the normal reinforced lead sheath of the submarine cable. Full-scale tests have been carried out on prototypes, including fatigue-bending tests intended to reproduce the maximum mechanical stresses anticipated. The answers were judged satisfactory. So technology, designed for nuclear power, is moving towards solving one of the problems associated with wave energy which would have been judged insuperable only a few years ago. It is of interest to note here that Mr. Glendenning who is no admirer of the Salter Duck, believes that with the help of Pirelli this will provide no obstacle, although he, of course, prefers that it should be used with Cockerell's Raft or preferably an air-turbine system.

I have perhaps laboured this point because it is essential to grasp that we are witnessing the development of a new technology in which the questions and answers are changing at very great speed. It is only in the last two years that the question of transmitting power from wave energy generators has been raised in a serious manner and investigated in detail by a body as earthy as the CEGB. We are present at a moment when practical people are grappling with the stuff that, until 1976, were dreams.

To return to Mr. Bott: in March, 1975, he gave the first major address on wave energy to the Royal Society of Arts. Indeed, it was probably the first time that a scientific audience of that eminence had been made aware of the subject in detail. He began by explaining how the waves convey energy and it is interesting to note that the basic facts that he spelt out would have been new at that stage to a large number of our most distinguished scientists who had never before had occasion to pay much attention to what Mr. Bott rightly called "a neglected corner of natural science". Had he been talking in this way of, say, steam power it would have been regarded as insulting to explain how that fellow Watt managed to do it; but on the waves, we were nearly all in 1975 at fifth form level.

Mr. Bott produced a table to demonstrate a fact which we have already encountered: that the height of the wave is the most significant measurement. Thus, a wave 1.5 m high with a wavelength of 15 m will produce 4.33 kW while a wave of the same length but double the height (3 m) will produce 17.9 kW, more than four times as much energy because the height is squared in the equation accepted by the scientists as depicting wave energy. But a wave of 1.5 m in height with a wavelength of 30 m will produce 8.9 kW, while a wave of the same height and double the length 60 m will produce only twice as much power — 17.8 kW. And when we move into the realm of higher waves, the difference is even more startling. A wave 6 m high will contain energy equivalent to 220 kW while one of 12 m will be 880 kW. That is why the highest waves are the most attractive, although they create the greatest design and construction problems.

This is the area in which hard decisions are soon going to be demanded: how much do you spend on building a device which can withstand, absorb and process giant waves and, if successful, will produce more electricity than a cheaper device in calmer water? At what point is it sensible to concentrate merely on surviving the really big waves, rather than absorbing them? Where does one strike a balance between the maximum energy concept of a plant based in rough, open sea against a rather less energy-intensive plant on the shore where a survival capacity of virtually 100% can be achieved? For Walton Bott, with his civil engineering background and knowledge of the sea, the constructional, operational and survival problem is naturally uppermost while other engineers, from other disciplines, will emphasise the significance of output and the advantages of a floating structure. It is like the battle at the turn of the century between the protagonists of alternating and direct current. That is one reason why the present stage is so exciting: we are privileged to be witnessing what must be, for the moment, a theoretical discussion on a practical issue that concerns all our futures.

Mr. Bott, having decided that the idea of placing operational plant in the open sea was "out of the question", turned instead to the shore. He realised that the fringing reef provided a stable foundation on which to build an impounding wall. The most expensive part of any sea barrier is the foundation and in Mauritius, as he put it, "nature has done it for us", thus greatly reducing the civil engineering cost.

All that would then be needed would be to build two crossbunds, at right angles to the outer wall, and turn the shore into an enclosed lagoon inside which the water would be trapped at a top water level of between 2 and 3 m above sea level. Low-head turbines and generators would be built into the crossbund walls and they would be turned by the flow of sea water rolling back into the ocean. Would the reef be able to take the strain? It would have to support a massive concrete ramp weighing many tons per foot run of its length, with the additional kinetic pressure provided by the sea. It would be a giant breakwater but with exactly the opposite function − instead of repelling as much of the sea as possible, it would have to allow waves over the top offering the minimum resistance above a certain height.

Mr. Bott's reply was: "Nature herself has built the reefs in the very teeth of the waves, even to the extent of providing an extra tough species of coral at the leading edge where the impact is greatest. This species (Madrepora) is in the form of a steel hard boss and has proved itself capable of resisting anything which the sea can hand out." (Personally, I would prefer to put that point less romantically and say that the only type of coral which could have survived under those conditions would be the toughest; this is a philosophical parenthesis rather than an engineering quibble.)

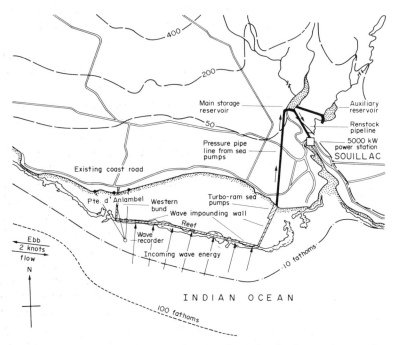

Figure 6.1   The Bott plan to receive sea water into a low-level reservoir, pump it up to a high-level reservoir with turbo-ram pumps and let it run down into a 5,000 kW power station.

The Mauritius team decided to experiment with small ramps but they could not be kept in place. It then decided to use wave recorders anchored off shore and to relate this to energy in tank experiments. So it asked the Ministry of Overseas Development for research funds and then, with the collaboration of the Crown Agents, arranged for the Hydraulics Research Station to carry out model tests in their tanks. Mr. Bott is filled with praise for their work. They provided all the information that he needed to arrive at the relation between wave height and period and retained energy after overspill. It varied the wave periods from six to 10 seconds, which are the normal limits off Mauritius. The period is significant because the arrival of each new wave tops up the level of the reservoir and changes the pattern of overspill back into the sea. Then the experimenters varied the wave heights and lengths, the wall heights, the depth of water immediately outside the wall and even the type of surface of the mock seabed.

A wall that was too high would provide a deep reservoir part of the time but would admit little water at other times and this could make wave energy a non-firm source of power. Mr. Bott wanted his scheme to be a basic, steady provider and that had to mean a lower height than a wall that would provide the deepest reservoir on favourable occasions.

Yet the lower the height of the wall, the smaller the reservoir would be and this would also mean a low turbine head as the water ran back into the sea. The cost would be greater — a head of 10 m for a turbine is about one-third the cost for a head of two metres. Instead of being able to provide two 2,500 kW sets, they would have had to think in terms of five 1,000 kW sets with all their auxiliary gear. Again, a low wall would mean a shallow reservoir and he calculated that this meant that the water power would be able to provide less than one hour's running at full load of 5,000 kW. Even this would need a dredged channel to enable the water to run out freely and that would cost more money. The power would be non-firm and there would have to be a thermal generating plant as an insurance. So everything indicated a higher wall, a deeper basin and a more economic head for the turbine — but with less potential energy arriving in the lagoon.

Mr. Bott calls it a Charybdis and Scylla situation; perhaps it would be more accurate to say that he was trapped between the devil and the shallow blue sea. The resources of the Hydraulics Research Station could help but could not solve the problem. While the discussion went on, in 1966, the price of oil started to fall. Authority looked at the scheme and said: forget it. As a practical proposition, it was dead, but Mr. Bott did not abandon the plan and it was much later that he hit upon a solution. It was to regard the water in the impounding basin not as the direct link between the sea and a turbine, but as an intermediate store. It would have a low surrounding wall which would enable it to receive energy for 24 hours a day. But then the captured water would be converted from low grade (so far as the hydraulic head was concerned) into high grade energy by pumping the water uphill, using the free energy of the sea to do it.

He devised a pump which was a rotary form of the old-fashioned hydraulic ram, which with its familiar thud-thud has been a long-established and reliable method of pumping water on farms and in villages. His pump combines the action of a water turbine and a water pump in a single rotor housed in one casing. In Mauritius, it would be designed to accept large volumes of sea water from the reservoir on its way back down into the sea. As the water flowed, it would drive a pump which would push the rest of the water uphill. The water going into the sea would drive a turbine. The water returning from the uphill reservoir would do the same, but from a much greater height.

A similar device was used more than 50 years ago in Germany when a "transformer pump" with a double runner was built on a weir on the River Izar near Munich. It was given its name because it "transformed" low pressure water into high-head pressure water. The mechanical efficiency is estimated at only 60% but this is of only academic interest when the energy driving the pump is free. It should be noted that our own CEGB uses pumped storage by feeding the upper reservoirs from power stations driven by fossil fuels, providing four units of electricity to push the water uphill for every three units produced by the water coming back. What goes up doesn't necessarily come down. Gravity absorbs 25%.

Mr. Bott points out that the storage required would be much smaller than is normally needed for a conventional river hydro project, where water flow can be insufficient for six months of the year. In Mauritius, it would be as little as six weeks in total, spaced out at different periods of the year.

As we know, it did not happen for the ludicrous reason that oil was then cheaper. But the plan remained and it was not forgotten at the HRS where the director, Robert Russell, acknowledges his indebtedness to Mr. Bott. "Our system goes on from the Mauritius scheme", he told me.

The scheme has been the least-publicised for two reasons: it is not photogenic and it has an unfriendly-sounding name, the HRS Rectifier. It was originally the Russell Rectifier but Mr. Russell, for reasons of modesty, preferred to keep his own name out of it. So, apart from the lack of attraction of a series of initials which need to be explained, the layman needs to be told what a rectifier means to an engineer. It sounds vaguely like a course of remedial physiotherapy. In reality, it is a method to rectify, or change, a movement. Thus, in the Oscillating Water Column, which we shall be examining later, the movement of air, which goes in two directions as it is sucked in and pushed out of a device, is "rectified" by valves so that it hits a turbine from one direction only. In the HRS Rectifier, the movement of the waves is changed from up and down into a stream of water which drives a turbine.

A better name would be the Russell Lock because it does resemble a lock gate dividing a high-level reservoir from one lower down.

The plan is to have a rectangular box standing on the seabed in some

Figure 6.2   The HRS Rectifier. Gates (which are non-return valves, like letterboxes at right angles) open and close as the water pressure changes and the waves roll in. The crest of the waves, and the increased hydrostatic pressure below the surface as the depth (and therefore the pressure) increases, force the valves to open. The water enters high-level reservoirs and finds its way out through the only exit, into a low-level reservoir, driving a turbine as it goes. It is sucked back into the sea as the waves fall, the pressure is reduced and the alternative set of gates opens.

15-20 m of water, possibly 5-10 km from the shore. It will be the size of a giant tanker. One half of it is divided into a column of gates made from concrete, with hinges of reinforced rubber. These gates open only inwards and the easiest comparison is with a letter box standing at right angles to its normal position. As the waves rise, the pressure increases and the water is forced into the compartment inside, which is the high-level reservoir. The pressure comes from the movement of the wave energy at the crest; lower down, it is a hydrostatic pressure — that is, the increased depth as the waves rise produces increased pressure and forces open the gates.

The gates are, like the letter box, a non-return valve and the water is trapped inside a tall box. As the water mounts, it is driven to the only available exit, a turbine which leads into the lower-level reservoir. And as the water surges through, the turbine turns, driving a generator and producing electricity. And as the water builds up inside the low-level reservoir, pressure increases on an adjoining series of gates, this time designed to allow water out but not to admit any.

Can such a structure stand on the seabed without being smashed to pulp? Those who are sceptical of its chances tend to say, "Well, if Russell says it can, then it can". His reputation in the wave industry is very high indeed. His team are probably unrivalled in their experience of making things stand up in the water.

It is a scheme which, as Mr. Russell put it to me, "has the virtue of extreme simplicity". That may also be one reason that it has attracted less attention than some of its rivals. It is probably also the most promising scheme for countries with a low tidal range, such as Mauritius itself and the Mediterranean. It has a major advantage over the original Bott plan, in that it uses the trough of the wave to suck water out of the low-level reservoir; his plan depended on potential energy responding to gravity.

The full-scale model would be designed to produce 10 MW on the basis of 70 kW/m. A 1/30th scale model is functioning at the HRS at Wallingford and the next stage planned is 1/4 scale to put to sea. An alternative, which would be more impressive, would be to build a set of flap gates full-scale, mount them on a "rig" — which means any test device that can stand in the sea — and test them in real-life conditions. This would make very good sense because the only part of the structure which involves new technology is the gates.

Mr. Russell is confident that they could withstand the pressure and is unusual among the experimenters in that he puts great store by that natural, old-fashioned material, rubber. "We would use stiffened rubber, incorporating something like tyre fabric" he said. "Rubber is infinitely long living in salt water if you keep it out of the sun. Keep it wet and keep it cool. You can find tyres on the beach 20 or 30 years old."

The thinking is typical of Mr. Russell's style. He is a tall, thin, aloof man with a rather sad manner. The mansion in the Oxfordshire countryside where his headquarters are based seems a generation away from the brasher laboratories at Marchwood and East Kilbride. One feels that Mr. Russell, though he is much too polite to say it, has a slight disdain for the newer arrivals on the scene. His Hydraulics Research Station has been advising chaps about building things in the sea for a long time now. . . . He also makes a natural ally for Mr. Bott, who is a kindly, elderly engineer; both are very far from the whizz-kid atmosphere that surrounds the Salter project.

The Mauritius and HRS projects have one asset that is enjoyed by none of the others: both are designed to form artificial lakes of huge area in the sea. They will be ideal for fish farming. There would be a continuous throughput of sea water, excellent oxygenation because of the wave and spray action, continuous throughput of plankton and other small fish organisms and a fully-enclosed sea area which could be sub-divided for the separation of fish varieties and sizes by means of floating sea cages, without the danger for the fishermen of working in the open sea.

There is another advantage, which could apply also to some of the other devices, which must not be overlooked. It is a word which can, and should, and does, make the oil companies worried: electrolysis. By passing a direct current through a liquid, it can be separated into its constituents. With water, it can be turned back into hydrogen and oxygen. The hydrogen can be used as a gas for cooking and other purposes or it can be combined with limestone and made into methanol, a liquid fuel on which cars and aeroplanes can run. The difficulty until now has been that the energy needed to produce the electric current is greater than the energy produced by the hydrogen. It is, therefore, wasteful. There is no sense in using up more fossil fuel to produce energy than the substitute fossil fuel will produce. But this ceases to be a factor when, in Salter's words, "the gods pay for the waves". We could well be on the eve of a development when we can produce not only electricity from the sea but also the petrol

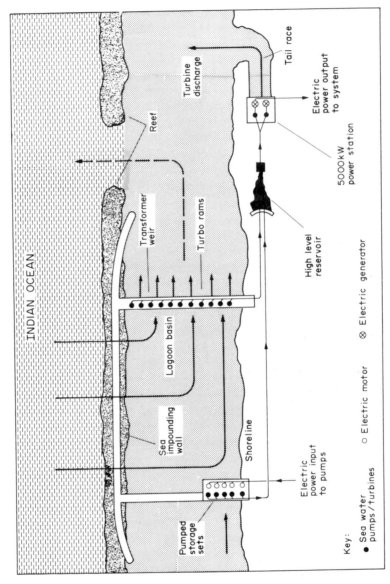

Figure 6.3   Schematic plan of idealised sea-wave energy project.

which we are consuming at a frightening rate by eating into our oil stocks. We may, in fact, have more than one tiger in our tank. We could also run de-salination plants on the "free" energy provided by the waves. Mr. Bott's conclusion is that world interest is mounting into the desirability of creating nucleus communities, virtually self-sufficient in energy, food, water and other essentials (and petrol has become one of them). A Mauritius-type project would, in his opinion, be "an almost perfect catalyst for such development".

Fish farming, de-salination and electrolysis are not the main purposes of wave energy. They are the possible spin-offs. All of them give a marginal advantage to the Mauritius-type project, or the HRS Rectifier or Russell Lock if one prefers that name. If only for these reasons the Government ought to be paying at least as much attention to what is being done in the quiet Oxfordshire countryside as it is to the more spectacular activities elsewhere.

CHAPTER 7

The Ship with a Broken Back

Sir Christopher Cockerell has already contributed a word to the language: he is the inventor of the Hovercraft. He gives the appearance of being the archetype of country gentleman who rages against the civil service and adores private enterprise. Yet he is also the person who first put forward the most revolutionary idea about wave energy. He first suggested that there should be "a crash programme" which would provide employment for our shipbuilding industry from public funds. He made the suggestion in a letter to *The Times* which practically nobody noticed; certainly none of the major figures in wave energy saw it. It appeared on December 21, 1976, a time when even engineers and scientists are too busy thinking about Christmas presents to read the papers properly, which goes to show that timing is everything.

He wrote:

> "We have a small, leisurely, safe programme for the development of some of the renewable sources of energy, with a 30-year target; meanwhile, each day we hear that the shipbuilding industry, or the oil rig builders, or the builders of conventional power stations are without enough work. Is there not a case for mounting a crash programme for the quick development of solar, wind and wave devices so that we can the sooner get work back into these valuable industries and save some of the misery and unproductive expense of further lasting unemployment? Consider the amazing crash programme for the Spitfire. It could be done again if we have the will to do it. 'There is a tide in the affairs of men which, taken at the flood, leads on to fortune.' "

Sir Christopher, lacking Salter's talent as a publicist, may have himself missed the tide with the timing of his suggestion but he did win the race to launch a fair-sized wave energy device. On April 19, 1978 he unveiled his Raft on 1/10th scale in the Solent and has ever since been producing an average of one kilowatt of electricity from waves roughly one-tenth the size of those expected in the open sea. The full-scale generator, roughly 100 m long, would produce a steady two megawatts and a series of rafts

62

stretching over 15 miles would be the equivalent of a 500 MW power station.

Sir Christopher received me in his home overlooking Southampton Water and explained his involvement in wave energy as a series of reverse operations. The waves had proved a problem for the Hovercraft and in seeking to overcome the damage that they could do he had realised their strength. He had thought about throwing a stone in the water and watching the ripples run out in growing circles and had contemplated the power that one stone produced and had wondered about the opposite configuration when the waves were rolling in towards a single object. And then he had developed a device which, he said, "is like a ship that has broken its back. The naval architect goes to school to find out how to design a ship that won't do that. Now we have to reverse the drawings."

Figure 7.1 Sir Christopher Cockerell.

The device, known as the Cockerell Raft, is actually a series of Rafts, or pontoons, linked by hinges. For convenience, everyone is calling it "a" Raft. The first idea was that a string of seven pontoons, linked by six hinges, would make one unit. Brian Count, a physicist at the CEGB's Marchwood Laboratories, close to Cockerell's headquarters, was able to show by theoretical work that fewer Rafts would be more fruitful. The unit has now been brought down to a line of three and may eventually be two. The units would be multiplied again and again, making a line of anything up to 1,000 km.

The Raft follows the contours of the waves. The front pontoon bobs up and down pretty freely as it is hit by the violence of the sea. The second one moves less, because the water has been flattened by its impact with the first and much of its energy has been absorbed. The third in the series is twice the length of the other two and is relatively stable. So you have a trio of pontoons moving in different phases and the difference is the motive force.

On top of each hinge there are two hydraulic jacks — long pistons inside cylinders. The movement of the sea drives the pistons up and down and sends hydraulic fluid moving inside a sealed system. The fluid is driven through a non-return valve into a manifold. This fluid comes from four directions. There is low-pressure fluid which flows out of a reservoir and high-pressure fluid from the business end of the pistons. It is mixed in the manifold and directed out of a single exit into a hydraulic motor. This is the opposite of a hydraulic pump, which is more familiar to most of us and which throws up a stream of fluid. With the hydraulic motor, it is hit by the fluid which drives it around. The motor drives a generator and the electricity sparks.

Sir Christopher is not enthusiastic about hydraulic power. "The kind of chaps you send out to maintain this won't be highly qualified hydraulic engineers in white coats," he told me. "So we want to get a system with the same advantages as the old steam engines — easy to maintain, whether it's sitting off Brazil or Scotland. You go aboard with a bloody great spanner and that is all you need. With hydraulics, the tiniest bit of grit puts them out. It is for me the practical things that matter. So you use the lowest form of technology that you can get away with."

He was touching on a point which has still to be decided. His own instinct is for a mechanical system and, like many of us, he thinks that the

Figure 7.2   The ship with a broken back. Cockerell's 1/10th scale raft working in the Solent.

fashion for hydraulic power systems is being overdone. In old-fashioned motor cars, it was certainly easier to deal with a bent brake rod than it is to bleed a hydraulic system nowadays. Against this, there is the problem of over-heating. With the enormous electricity generators that we are considering, a mechanical engine would create problems.

It would require something like a row of cooling towers on top of the Rafts to keep down the engine temperature. On the present models, the hydraulic fluid is cooled in a reservoir containing three times the capacity of the system but for the full-scale Rafts it would need to be a much larger vessel, a swimming pool in size, and this again would add to the structural difficulties. So the tentative solution being pursued is to use the sea water itself as the fluid, dumping it over the side when it gets hot and replacing it with cold water.

One of Sir Christopher's engineers, Michael Urwin, elaborated. "Hydraulic fluid is much stiffer, like brake fluid, than water and that is why we are using it on the model. But in the real-sea situation, with full-scale generators, a small loss of efficiency doesn't matter. In fact, by going for a very efficient system, we would get more power in a rough sea than we could cope with so a less efficient system is in some ways an advantage."

Sir Christopher is keeping an open mind on what form of power station he will settle for. "We have been looking hard at pumps on top", he said, "and there are pros and cons. You have got to take into account maintenance. Can you do it better right inside the device, in a sort of ship's engine room? Or could you just put a chicken house over the pump?"

On the model, the working parts are on top. This is inevitable for one practical reason: the Rafts are too small for a man to get inside an "engine room". The likelihood is that the full-scale Raft will have an enclosed room, partly beneath the surface, where maintenance and repairs can be done in comfort, apart from the rolling motion created by the sea. It will be necessary to have helicopters available, either on a nearby shore or perhaps on the deck of the Raft, ready to pull off the engineers, fast, when a storm is approaching.

One problem which seemed for a time to be a major one is proving less worrying. This is the fear that in a strong sea the front Raft would be thrown right over and would cease to function. Tests have now shown that the hydraulic system serves to dampen the motion of the front pontoon, acting rather like a shock absorber, and the fear that strong seas would take out whole lines of Rafts seems now less worrying.

For the Raft, as for all the devices, mooring is a major problem. The Solent model has been held down with four flexible ropes. They need greater elasticity than in the deep sea because there is a rise and fall of three metres in the tide, in water only 8–10 m deep. This problem will be less in the deep sea. The ropes, 50 m long, go down to a block of concrete which weighs 136 kg in the air and which itself rises and falls with the tide. The block is linked by 10 m of chain to a 180 kg anchor. The main purpose of the mooring is to hold the Raft on station; its own weight, and particularly the weight of the third and largest pontoon in the series, holds it down on the surface, and provides the difference in movement of the three pontoons and thereby produces the energy.

Mr. Urwin said that their plan was to go about 50 km out from the shore where the long, rolling waves would suit the Raft. In shallow water, he noted, the height of the waves increases and the length decreases. The Raft is most efficient when the wavelength is the same as the length of the Raft. So they would most like waves of 100 m — those with a period of roughly eight seconds, and a height of 10–12 m. With such waves, the Raft would be 100% efficient though there would still be a loss in generation and transmission — possibly 10% for each.

With most longer waves, efficiency would decrease but the output would remain the same. But with the very long waves, the whole unit of two or three would tend to ride in one section, instead of moving at different paces and for different distances. In that (rare) case, no energy at all would be extracted because the three pontoons would be just one bobbing Raft with the hinges open.

The problem for this, as for all wave energy generators, is survival in a really big sea. Mr. Jim Platts, one of the senior engineers on the project, put it this way: "The average sea gives you the value. The extreme sea gives you the cost. If you can find a sea where the average is low but the extreme is even lower, it might prove cheaper." He estimates that when the sea is producing more than 50 kW/m, the main interest is in survival because it is not going to be possible to absorb and process the really gigantic waves. Off the Hebrides, it is possible to have waves producing 1,000 kW for a day and 10,000 kW for a minute. No-one has even

Figure 7.3   How Cockerell's Rafts will look in real life. An artist's impression of a trio of pontoons riding in waves which lift the smaller, front pontoon while the second and third in the series stay relatively stable 8–16 km off shore.

contemplated designing a device that could handle such enormous waves.

One advantage of the Raft is that it is designed to allow the big waves to "slop over the top" as Mr. Grove-Palmer, secretary of the Wave Energy Steering Committee, puts it. He compares it with the Kon Tiki which has no freeboard — the part of a ship's hull reaching up from the waterline.

None of the other devices has this advantage, although the Duck comes close to it. The exact opposite is the HRS Rectifier which will cope best with local seas — the sort of excited condition that you find closer to shore where the device would sit on the bottom and enjoy the benefit of waves which were being compressed by the rise in the seabed as they approached the land. These waves have a shorter length but a greater height. The Cockerell team compare the HRS Rectifier to a high-rise block of flats standing in the water. Their own lines of Rafts would be more like a sprawling estate of semi-detached, two-storey houses.

The more one sees of the different devices, the more likely it seems that different seas will provide the setting for different devices, with the Salter Duck probably facing the roughest seas off the Hebrides, the Cockerell Raft coping with middle seas and the HRS Rectifier standing close to the shore and receiving the livelier but less powerful waves; the Oscillating Water Column, which can be used in the open sea from the size of an 11 m high navigation buoy to a giant structure weighing perhaps 20,000 tons, might prove the device most adaptable to different conditions.

It is perhaps not an accident that Sir Christopher was the first to get electricity out of the (fairly) open sea around Britain. He has a streak of the old buccaneering engineer. He first went to see Mr. Goodwin at the Department of Energy, but that was "before the Arabs upped the price of oil and it didn't look as though the thing was viable". But he pressed on, using his own money, and still expresses a little indignation over the fact that Salter had the facilities of a university and Government backing before the official grant was made. Sir Christopher went to see the British Hovercraft Corporation and they "out of the kindness of their heart" agreed to make some models and put them in a tank. Then he interested some friends: E. W. H. Gifford and Partners, civil and structural engineers based nearby in Southampton. From this informal start, the company called Wavepower was set up.

He talks staccato: "What is the replacement value of the present CEGB? £30,000 million or add another nought on that. The money is so vast that we should press ahead to a realistic stage on more than one of these gadgets. We don't limit ourselves to buying one motor car or one TV. Never is it one individual. You could not have a school of painters with only one painter" (Sir Christopher is a trustee of the National Portrait Gallery, worried that we are not doing sufficient to preserve paintings).

"With wave energy, there is an extraordinary advantage: we don't have to standardise at any time. With the railway system, you have to decide a gauge and whatever forever. With wave energy you could have a fleet of devices which could be one design, and you could have a Mark 2 next to it and a Mark 3 next to that. Because design can improve, you can change again and again. Eventually, you chuck out the early ones and replace them with Mark 7. That is one of the delightful things about it.

We know we would produce energy. The question is: will it survive? Will it break away? How much maintenance will it need? You are dealing with a probability. This is not static. It is living in a filthy environment. Not enough is known about it. Look at aeroplanes or washing machines. Designers have to have at least four shots before they have got a chance to have ironed out the worst of the trouble. The designers have got to have a chance of making mistakes. Think of corrosion and metal fatigue and storms and maintenance and cable connections and so on. We will make some errors. Everyone makes them. They built the M.1 and now you have to repair the thing. You must allow the engineer to have these shots.

It would be fatal to reduce the scheme to one team and one device. That is what the accountant would say, rationalise the thing. It is wrong. Not all of the good ideas will come out of one of these schemes. But civil servants are so naive. And this is what enables the accountant to muck it. The civil service chap must not blot his copybook. It is safer not to put your nose out too far. It is not the people I quarrel with. It is the set-up which produces a certain result which is not favourable to new projects."

He excludes the Wave Energy Steering Committee at Harwell, and its Technical Advisory Groups, from his attack: "One feels at home with Harwell. They want the job done well and as cheaply as possible. They have taken on a good bit of the attitude of private enterprise. They know that hanging around costs money. That is so much better than my previous experience with the more orthodox bodies. In the development of anything, if it is done well it must show a waste of money. The civil service can't justify it. It has not got a column which justifies it.

While the world still has fossil fuels, it is the equivalent of Mrs. Thatcher having a lot of groceries in her larder which she can call on when she needs them. When you are right out of fossil fuels, you cannot call on supplies you have not got. To use our oil to produce electricity is just terrible. We are using up the capital of the country. Anyone who lives by using up his capital is a mug. It does not matter whether the predictions about the length of time we have are wrong. When you run out, you have lost your money in the bank."

And then, with an echo of an old controversy, he turned on me and demanded: "When the scarcity starts, are we going to be so involved with the EEC that they demand half of our oil output? They have achieved half of our fishing already."

A splendid, charming old gentleman with the mind of a young radical.

CHAPTER 8

The Duck that won't lie down

The hero of this story is Stephen Salter. He is a good man and, in the words of the old Blues classic, a good man nowadays is hard to find. He is also a difficult man. The first time I visited him in his laboratory at Edinburgh University, in 1976, I had just encountered the idea of wave energy and it took him about one minute to realise that he was dealing with an ignoramus. He rattled off a few mathematical formulas, stared at me scathingly, handed me a bundle of documents and left me to his assistant, David Jeffrey, who was more tolerant. No-one would accuse Salter of tolerance. He is a tall, lean, intense engineer suffering from what Louis Aragon called the passion for the absolute; Sigmund Freud defined the condition as fanaticism. Salter needs to be that way. No-one without that mixture of imagination and determination could have persevered as he has done. In one of the documents that he gave me back in April 1976, he illuminated the whole subject with a marvellous phrase whose significance has still not been grasped by some of the people engaged in quantifying the project: "Efficiency itself is of no concern when the gods pay for the waves". It ought to be written up on the walls of the CEGB's headquarters.

He also maintains a sense of humour. Describing his early attempts to win wave energy, he remarks that "the obvious extraction mechanism was something like a lavatory ball-cock bobbing up and down". That captured about 15% of the available energy. Then, with the hinge below the surface, he captured 60%. He moved on to a vertical flap but this yielded only 40% because it was creating reflected waves by its own movement to the rear. He tried something shaped like the British Standards kite-mark and "its round rump" displaced no water as it moved and captured 70%. Finally, he devised the duck which can win 90%. And so it began.

His critics are numerous. As Mr. Goodwin said (see Chapter 5), Salter is inclined to make fools of people who had greater facilities than he enjoyed and "this is a way of influencing people but not of making friends". Salter has made enemies.

For instance, I have been told that he poses as "Dr." Salter, while lacking a doctorate. The explanation is that he was once referred to in a newspaper as Dr. Salter and newspapers and other sources have repeated the error. It is infuriating for many journalists to discover that when they describe him rightly as Mr. Salter, some helpful ass looks up the cuttings in the library, sees that he is called Dr. Salter and "corrects" the text in order to avoid downgrading and offending him. That's newspapers. Nothing is more difficult to erase than an error. But Salter is not responsible.

Again, I was told that he was nothing but an amateur with oily fingers and no academic qualifications. "He started as an apprentice and just got Edinburgh University to give him facilities", said one of his critics. Another remarked that he enjoyed an unfair advantage because he had the university's laboratories and youngsters who came in to help, unpaid, in their lunch breaks. The way to discover the facts, plainly, was to return to Edinburgh.

I rang Mr. Salter and this time the reception was different. "Ah", he said, "you wrote an article in the *Sunday Express*. What a pleasure it was to see something in a newspaper that was accurate." From then on, we could talk.

He did, in fact, start as an apprentice — with British Hovercraft, Cockerell's firm on the Isle of Wight. "I did start at the very bottom", he said, "learning how to use a file and they eventually taught me as much as Cambridge". Which is where he went on to, to take a physics degree. He had the essential qualification of A-level maths which enabled him to become a professional engineer. (This, in passing, is the hurdle which everyone has to surmount. We are producing school-leavers who have given up maths at O-level and can never return to any branch of technology with a hope of a useful degree).

Some of the sniping at Salter's qualifications may arise from a misunderstanding: Cambridge awards its science degrees as B.A. and Salter went on to take an M.A. but a superficial reading of his qualifications might suggest that he was just another English Literature fellow.

He has surrounded himself with a team of enthusiasts ranging in age from 17 to 34. One of them, Glen Keller, is an American who was studying ocean engineering at the Massachusetts Institute of Technology

Figure 8.1 The new Edinburgh wave simulation tank in action. The photograph shows one particular pattern of waves that can be created in the 30 m x 12 m tank. This crested "freak wave" at the centre of the tank would, at full-scale, measure more than 60 m from peak to trough.

when Salter gave a visiting lecture. He asked if he could join the Edinburgh project after graduating, wrote to Salter a year later and was accepted. Salter insists on this form of approach. He does not *seek* recruits because he wants people who want to work with him. He once said that his ideal collaborator should be an expert in physics, electronics, naval architecture, stress analysis, biology (because of the barnacles), computer programming, meteorology, economics — and public relations. In the event, the public relations were handled with natural talent by Salter himself. His recruiting method and his own personality have created a team with striking enthusiasm. The secretary, Miss Jean Richmond, is as immersed in the waves as any engineer and when they move in a group to the canteen for lunch the work is not interrupted. They talk shop.

They have built a testing tank, 30 m by 12 m, made of concrete and glass with miles of Dexion on top. It was completed on Xmas Day, 1977, with Salter (and his wife) present as the final nuts were tightened. It cost £100,000 and holds 100,000 gallons of water — £1 a gallon, just like petrol will soon be if we do not increase our energy resources.*

It is the most advanced testing tank in the world for the waves. They were told that it would be impossible to build. Along one side are 89 wave makers — yellow paddles, each with its own control panel, designed by Mr. Jeffrey. The controls push the paddles forward at varying strengths. The control panels are stationed up above each paddle and each one contains 14 integrated circuits, all of which had to be wired and soldered by hand, together with resistors and capacitors. Each panel looks like — and indeed, is — a delicate electronic printed circuit board such as are mass-produced for transistor radios and stereo amplifiers and TV sets. But the wave controls each needed individual assembly. It meant 1,246 integrated circuits put together with delicate fingers and sharp eyesight by a man and a boy: Mr. Jeffrey and Ian Young, a student of 17. It took them two months of non-stop dedication.

The effect of the work is fascinating to watch. The operator presses a series of numbers on a keyboard, or feeds in a length of tape which has been cut in advance with a pattern of holes, just like the tape that is fed into a teleprinter, for the more complicated operations.

Now imagine this huge swimming pool slowly being crossed by a majestic swell, a long line of rolling waves maybe 8 seconds apart as

*At the time, petrol cost about 70p a gallon.

though they had travelled across a fetch of 160 km with a wind speed of 30 knots behind them. But in a normal swimming pool they would hit the opposite side and bounce back as reflected waves (or like echoes if they were sound waves). In real life, the waves end up on a beach, which absorbs their energy in a wasted froth of breakers. It is obviously necessary in a wave tank to create an artificial beach and this has been done with weld mesh. It looks like a series of pillars of steel wool, such as one uses for scouring dishes. They are about 2 m high and they soak up the energy to ensure that the waves go forwards but not back. The beach takes a tremendous pounding and it is startling to realise that once a line of ducks, at only 1/150th scale, are sitting in that storm sea, the water will be a dead flat calm behind them — as Salter has already demonstrated in narrower tanks.

But the real sea never performs in such a regular way. Waves come from different directions, with different periods, overtaking one another and crossing one another's path, as Mr. Draper explained in Chapter 2. The question then that has to be answered is what will happen to real-life wave-energy generators when they encounter a turbulent sea with, literally, an infinite variety of pressures that must be absorbed and processed? It is here that the controls begin to show their worth. They receive different instructions from the computer and operate the paddles at varying pressures and speeds. Within seconds, the sea can be turned from a gentle swell into a fury of turbulence. Waves criss-cross, smash into the "beach", form humps which glide across the surface in a diagonal pattern. The Salter team claim that they can emulate any wave spectrum in any sea.

This gives them two separate results: they can see how the energy varies with frequency, and they can see how the Ducks will perform in a sea which can produce the equivalent of 500 kW/m — "the sort of sea in which we want to survive", as Mr. Keller put it. They have built a spine of PVC with ping-pong balls in clusters at each joint. They plan to put between 20 and 30 Ducks at 1/150th scale on the spine and see how they manage.

I have laboured what may seem like a secondary aspect of laboratory testing for one reason above all: the tank, the size of a large swimming pool, with waves raging across it and rising above "sea" level in just the way that waves do at sea, is intended for testing a string of Ducks of only

ENERGY FROM THE WAVES

Figure 8.2   The picture that revolutionised the thinking of the Department of Energy: a wave glides into the nose of a 1/150th scale model of a Duck and a calm stretch of water emerges at the other side. Stephen Salter has captured wave energy. The waves are being absorbed with an efficiency of about 80% and very little power is escaping to the left, where the water line follows closely the line of the top of the glass tank.

1/150th scale. That is to say, the diameter of the spine on which they pivot will be a mere 10 cm. The larger models which are being tested on Loch Ness are 1/15th scale and their diameter is one metre. The real Ducks out at sea will be about 15 m in diameter. That, remember, is the diameter of the spine. Then you have to add the Duck's beak, made out of concrete, which bobs up and down in the waves and you get a device measuring from the Duck's beak to the Duck's arse probably more than 30 m – that is, a building the size of an eight-storey block of flats.

And if it is to operate well, it will have to be one of a string of about 20–30 Ducks on one spine, with spaces between each Duck. So one wave energy device would probably be about 1,200 m in length, nearly a mile. And that should produce between 30 and 50 MW. Think back to the swimming pool, multiply by 150 to get the diameter of the spine, add the beak, then multiply by 30 to get the length of 30 Ducks that you will probably need and you have some idea of the size of the beast we are talking about. And then you will start to understand why it needed someone close to the frontiers of genius and lunacy to persevere and to persuade the Civil Service that he was talking sense.

Stephen Salter has retained his confidence and his coolness through an extraordinary odyssey. He does not try to brush aside the difficulties. For instance, he readily agrees with his critics that the Duck does not lend itself to the "penny pieces" idea that is championed by Sir Christopher Cockerell and many others. It would not be possible to station *one* Duck in the sea, in the way that you could experiment with the Raft and the Oscillating Water Column. The Duck must be built in a string of 20 or 30, so that the pressures on the spine are evened out, as different waves hit different Ducks. Against this, Salter's team have two impressive arguments. They believe that their device will be able to cope with the worst forces of the sea. It has the most ambitious design and is consequently the most difficult to build and maintain. But it is designed to be hit by furious waves and survive because the biggest part of it will be the moving part. It is the spine which has to stay fairly stable while the Ducks will bounce up and down. With the Raft, exactly the opposite applies: the third of the trio of pontoons is twice the size of each of the other two and it is the big one which has to stay steady. And with the OWC, the whole pillar of concrete needs to be steady in order to create the maximum relative motion of the air bubble. So one can contemplate a situation in which the Ducks could survive off the Outer Hebrides and, although they would present problems in construction and servicing, would be very fruitful sources of power while other devices would survive more happily in calmer seas, producing less electricity but costing less to build and maintain.

The second advantage is that the Duck would be more easily mass-produced than any other because it comes in smaller, more numerous units. One OWC is equal to, say, a string of 15 Ducks (and three Rafts). The enormous tower of the OWC would drive one turbine. Each Duck would drive perhaps six or 10 pumps, which would drive a turbine from each Duck. There will be nothing like the pillar of concrete standing up to the sea's assault.

One of Salter's team put it in a striking way. A conventional power station, he argued, could be regarded as costing £400-500 per kW. A motor car engine, if used in combination with a generator, could produce easily 20 kW. The engine would cost only about £50 because of mass production. So the cost per kW would be £2.50p. He was satirising the problem. But, as with all good satire, it contains the nucleus of truth.

How will the Duck work? First there is a spine with a diameter of 15 m, and a length of roughly a mile. Most of the structure will be built of reinforced concrete. There will be hydraulic rotary pumps based inside the spine and the aim is to make the spine as stable as possible while the Ducks flop up and down. The power is the relative motion between the Ducks and the spine. If the spine yields too much, you lose power. If it did not yield at all, it would break. As Glen Keller put it, "all the Ducks are trying to make the spine move as the waves hit them. You want to make the spine long enough, and with sufficient number of Ducks, so that when you average all the Duck torques on the spine, it should come close to zero." Rather like two chaps of equal strength clasping hands, putting their elbows on the bar counter and pushing against one another, so that their hands are immobile, except that each individual Duck will be driving a pump inside the spine. That pump, the front end of the system, will be pushing out fluid, either water or oil, inside an enclosed system. The fluid will drive a generator which will be installed in a black box, probably stationed between the Ducks and certainly above the surface because, as Mr. Keller put it, it would be "very painful" to have to service it inside the spine under the surface of the water. Or the fluid might be pumped ashore to a land-based generator.

Figure 8.3   Salter's Ducks. A string of Ducks, held together on a spine, bob up and down at different moments as the waves hit them. The variation in timing helps to even out the strain on the spine. The bobbing movement of the Ducks drives pumps which, in turn, motivate a generator and produce electricity.

They needed, he said, a system like the electric lights on a Christmas tree, to ensure that when one light (or Duck) went out, the others continued to function.

They had contemplated direct drive of a generator from the Ducks but decided that it would be too slow. It could be geared up but this would need 15 m gears which would depend on close tolerance, something that one did not wish to rely on.

The spine remained one of the key issues to be solved. In addition to the torques on the spine from the power take-off, there would be bending movements caused by the waves themselves and these would push the spine sideways and up and down. If the spine were entirely rigid, it would break, so it must have joints between sections. It would obviously be easier to build if it could be rigid; "Once it is flexible, it gets more subtle. We have to reach a compromise."

Mr. Richard Jefferys, an expert on hydro-dynamics, is particularly concerned with the problem of efficiency in different wave conditions, something which is of even more concern to the group experimenting at the National Engineering Laboratory on the Oscillating Water Column. From their standpoint, it is probably more "efficient" to take a smaller amount of energy from a wider range of waves than to be "100% efficient" in absorbing the energy of fewer waves. Indeed, with the OWC it is possible actually to have more than 100% efficiency by absorbing energy from different directions. But what is needed for all the devices is a means of accepting and transmitting as continuous a stream of energy as possible, even with lower "efficiency".

The way Mr. Jefferys regards it, so far as the Duck is concerned, is: "You could make a Duck that was very efficient for one particular period but for other periods it would drop off" (period referring to waves, arriving at anything between five and 15 second periods). "You can tune it so that it is better over a wider band of frequencies but is not quite so good at its peak. We are thinking of 100 kW/m as the maximum amount of power the Duck can handle. But if you got 2 MW/m hitting it, and it is 5% efficient in these conditions, you would still get 100 kW/m."

David Jeffrey, one of Salter's earliest collaborators, would like to see full-scale work on parts of the Duck. He wants to simulate the motion of the sea by "driving" a Duck with a hydraulic ram, so that they could see how the power fluctuates as the Duck wobbles. He would also like to try out drive mechanisms, how to communicate the force of the Duck's movement to the hydraulic pumps. He is impatient of the step-by-step progress that has been the fashion until now. "The only reason for the 1/10th or 1/15th scale work is to prove that the 1/150th scale models' forecasts are correct", he said. "There is no reason to say that you can't say precisely what will happen to full scale. On 1/150th, our predictions were still accurate for 1/15th and there is no reason to assume that they would not be correct at full scale. Once we have the 1/150th working on a spine, we could go up the scale without difficulty. We would like to test certain parts at full-scale. At 1/15th scale you could not model the power take-off. The reason is that you want to try to design as far as possible using things easily available off the shelf and small hydraulic systems are not in that category. They *are* available for hundreds of kilowatts but there are not enough diverse systems for dealing with hundreds of watts.

We are talking about power, and the power of the 1/15th scale model is going to be 3/10,000th of the full power. Three watts from 1/15th scale would mean 10 kW full-scale."

I had better butt in here with an explanation. Scale does not go up in a straight arithmetical progression. For instance, it is obvious that if you take a three-dimensional model and multiply it from 1/10th scale to full-scale, you do not simply come up with something 10 times as big. As every dimension is multiplied by 10, then a model of 2 m by 2 m by 2 m becomes 20 m by 20 m by 20 m; the difference is between 8 m³ and 8000 m³. In the same way, the power output is greatly increased. Mr. Jeffrey was hinting at something that has aroused a great deal of backstage fury in the wave energy circle.

The project for a trial on Loch Ness came not from Edinburgh but from Lanchester Polytechnic in Coventry. There is little love between the two scholarly establishments. And the Poly is critical of the National Engineering Laboratory which provided the power pack for the trials. Somewhere, someone blundered, as Tennyson would have said. And, as President Kennedy did say, defeat is an orphan. At the time of writing, the Loch Ness trial is more than six months behind schedule and the spine has broken once. Edinburgh would say that Lanchester should never have boasted that it would put out a string of Ducks by the autumn of 1977. Lanchester would retort that it would have succeeded in doing so, before the weather broke and the Loch froze over, had it not been for the NEL, which failed to provide the power pack in time. And the NEL says that it received a request only on September 29 to produce a special generator which was delivered on November 1 and was produced in the usual NEL style – which means that, as one of their spokesmen put it to me, "we put much more effort into these things than we charge". This I can confirm from my own knowledge of NEL. Its competence is unquestioned. Yet the Poly then decided that the generator needed modification.

A lot of people will be arguing about who was to blame. I am not going to jump into that whirlpool. I merely report that in Edinburgh it is being said with bitterness that the Poly jumped on to a bandwaggon by a side door. Edinburgh received a friendly communication offering to put the proposed design of the Duck through an analogue hybrid computer, to see if it was the most efficient design. The next thing that Edinburgh knew was that the design had been modified by Lanchester and models of

1/10th scale (as Lanchester sees it) or 1/15th scale (as Edinburgh sees it) were being built to the Lanchester design. It is different from the Salter Duck and Salter feels keenly about this. He says that he has never given it his sanction and yet stands to be blamed for any faults. He told me this *before* the spine had broken.

Yet Lanchester Poly's electrical and electronic engineering department, headed by Dr. Norman Bellamy, is convinced that it is making a great contribution. Dr. Bellamy, who comes from a mining family and started his working life as an underground miner, is highly enthusiastic about wave energy and is convinced that his department can make a great contribution. "I started out very sceptical", he said. "Then I realised that the energy was there. And it is there forever." He is hoping to move on from Loch Ness, once it is functioning properly, to 1/4th scale, analysing the data as he goes. He says that there are many months of analysis to be done on a computer.

The Salter team are cool to the whole concept of going up the scale in this way. They believe that what can be done at 1/150th scale, in their extraordinary tank, will prove accurate for the full-scale prototype. "The only reason for 1/10th scale work is to prove that the forecasts from 1/150th scale are correct" said one of them, commenting bitterly on Lanchester's contribution. Edinburgh would prefer to devote the main effort now to full-scale tests on certain parts of the device.

For the present, these conflicts would probably be described by a management consultant as "creative tensions". They remain creative so long as the operation moves forward. Yet problems of this type will increase as the project grows and the Wave Energy Steering Committee, which is allocating the money, will have to find a way of ensuring that if different centres are engaged on the same operation, they are tolerably harmonious. Distance is one problem. The huge generator from the NEL had to go backwards and forwards by road from East Kilbride near Glasgow to Coventry twice before it could be taken to Loch Ness. It would have been easier if it had gone from East Kilbride to Edinburgh and then up the road to the Loch. And, after all, Edinburgh University does have computers. . . . One of the great advantages of the Cockerell project has been the compactness of the contributors. Wavepower, the British Hovercraft Corporation, Gifford and Partners, Cockerell's own home and the Marchwood Laboratories of the CEGB are all close to Southampton and in constant contact. It makes life, and progress, easier.

One can see the point of trying to spread the Government's gift money but the WESC will soon have to decide whether it can maintain benign supervision from a distance or whether it would not be better to take a step backwards, allocate lump sums to the main centres, and leave them to work out the best way of dividing up the work under their own direct control. This will be a major problem when we start to deal not with £5.4 million but with hundreds and thousands of millions.

The incident will probably be seen in the future as only a trip wire and there is no reason why the Salter Duck should suffer from it, provided that the Department of Energy and Harwell draw the right conclusions. And these conclusions will be reflected in the next stage of development and how much money is available and how it is allocated.

The final word on the Duck belongs to Salter. He told me that for the next stage he would like to build a full-scale power take off and one joint. He is not pushing for an immediate full-scale spine and string of Ducks but, if national policy required it, if something nasty happened in the energy field, then he could do the job for £30 million. That would produce the equivalent of a 30 MW power station. It would be done without hesitation if nuclear bombs fell on the Persian Gulf and Saudi Arabia.

How much energy could the waves provide? Salter said that he thought they could produce 30% of our electricity. With respect, this is plainly thumb-sucking (and modesty) and no more significant than the 3% which is mentioned by some people around the Department of Energy. We could, if we needed to, multiply the lines of devices almost indefinitely. There are now cautious experts who say that we could have parallel lines only 80 km apart and, with the reduced fetch, they would operate at only 50% efficiency downwind. But the nation would still be in profit because, in Salter's words, "efficiency is of no concern when the gods pay for the waves".

I put to him what is the key question for the future: could wave energy, in his opinion, ever be a firm source of power? Yes, he said, he believed that it could be, if there were devices stationed in different oceans. Coming from someone whose claims are so unextravagant, this is a tremendously important statement.

The Government will shortly have to decide whether to give Salter £1 million for his "slice of Duck" or whether to make a really imaginative decision: an award of £30 million, little enough in the overall energy budget, could provide work for our shipyards and produce a string of 50 Ducks, the equivalent of a 30 MW power station. There is an excellent site available off Dounreay, on the Scottish north coast, where there is a nuclear power station with cables already existing to the grid, and the cables are being under-used. That is how the Japanese would be thinking. Is there anyone in Whitehall with the same spirit?

CHAPTER 9

Now We Copy the Japanese

The catalogue that came through my letterbox from Japan made it all real: for the first time, I was looking at graphs showing not what the various wave energy devices would or should be able to do, but what they had actually done. I understood just what Mr. Goodwin of the Department of Energy had meant when he talked of his discovery, "at a point where credibility mattered", that wave energy generators were working. We are talking about something no bigger than the dynamo on the rim of a bicycle tyre, compared with a power station needed to fuel a factory or a city. But it works.

The Japanese device was invented by Professor Yoshio Masuda, a former Naval Commander. It consists of an upturned canister, floating in the water, with two holes in the top. As the waves rise and fall inside it, air is sucked in and blown out. The stream of air drives an air turbine which, in turn, drives a generator and produces electricity. Any engineer will tell you that when you have a device which works at small scale, the problems that arise when you move upscale can rarely be anticipated in advance; but press your engineer and he will admit that the unforeseeable problems are invariably capable of solution. For months I have been looking at documents showing what can be done with various devices both in laboratory conditions and with scale models on fairly open water; it is difficult to convey how illuminating it is, at that point, to see what has happened when the wave energy generators have been functioning in the open sea, in real life, for a purpose, to guide shipping. This is the step from the theoretical to the practical.

How "efficient" is it? My answer is: who cares? The more precise reply is that, when fuelling navigation buoys, it is actually too efficient. According to Mr. Ernest Humphrey, who is responsible for engineering development at Trinity House, which is buying three of the Japanese buoys, the six 2 volt batteries which receive the electricity are "on charge almost all the time and if they reach a fully-charged condition they shut themselves off to prevent the water boiling away. The generator is designed to disconnect itself when the batteries are fully charged. Power is dumped off the side." The size of the batteries is convenient for the lay reader: it adds up to 12 volts, the same size as a car battery, which has to drive a starter motor, provide for short journeys, boost the heater and the windscreen wipers, as well as all the lights; the Masuda Buoy is needed only to light up a 60-watt bulb.

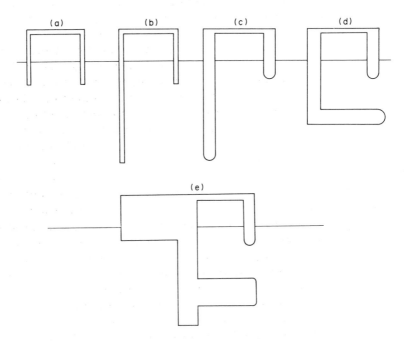

Figure 9.1   How the Oscillating Water Column (or Masuda Device) has evolved. (a) shows the original idea of an upturned canister; (b) shows the first improvement by the National Engineering Laboratory giving asymmetric design; (c) eliminates the sharp corners which cause high velocities and high flow losses; (d) with a base plate, provides the most efficient design. Then there is (e) the final version, which is a combination of (d) plus a space to house the turbines and other equipment.

As an additional boost, the Masuda Buoy can be fitted with a sun switch which ensures that the light goes on only when the sun goes down, an optional extra that can be useful in, for instance, summer time when the daylight is long and the nights are short and the sea may be calm.

Trinity House has had a sample working in the Irish Sea for three years, and 300 are functioning in the Pacific. Experience has shown that the batteries (divided into six units deliberately, to guard against failure) normally last for about three years – about 50% better than a car battery. The lights have twin filaments which come on separately and each one has an expectation of life of 12 months. So Trinity House plans to change the lamps every 20 months and thus gain a margin of four months before the light should fail. But it is still going cautiously. As Mr. Humphrey puts it, "We have done the initial trials but one needs to find out those things which we don't have a way of simulating. There are things like bumping the ship when it is laying the buoy and environmental conditions and human factors."

The buoys are to be manufactured in Britain by AGA Navigation Aids who are importing the generator and adapting their own standard buoys to carry it. The Japanese have used an anti-corrosive aluminium alloy. AGA's spokesman told me: "We will be one of the first to put it on a plastic base."

Does it not seem like a satire on our timidity, our lack of confidence, that our contribution to this revolution in the creation of a new form of energy should be – plastic?

It used to be accepted by us that the Japanese were great imitators. They copied all our best ideas, mass-produced them with cheap labour and flooded the world market. Indeed, among many people there was a phrase for the smallest discernible type sizes: we called it the "Made in Japan" type, when it was the trip item at the bottom of an insurance policy or an industrial agreement, the sub-clause that its authors were rather ashamed of. The phrase has tended to disappear, along with much of our electronics industry and our car manufacturers, and we need now to recognise that the Japanese have made a significant contribution to wave energy. Their need for fuel is greater than ours, as they have practically no natural resources of their own and are, for reasons that date back to Hiroshima and Nagasaki, not enthusiastic about nuclear power; indeed, they like to send us their nuclear waste, as we know. Yet circumstances have forced them to go nuclear and they are today the world's second largest nuclear power, having overtaken Britain with 14 nuclear power stations operating, 12 more due to come into operation in the next four years and another six in the planning stage. They are being forced to go nuclear or to find another source of energy; hence the urgency that they are putting into the main alternative.

Their brochure takes one into another world. They talk of producing 30 W in waves of 40 cm with a three-second period. This is a very small sea compared with the waves of perhaps 1,500 cm which we have become accustomed to thinking about in the North Sea and the Atlantic. The explanation is that buoys are usually used in calmer waters, particularly in narrow straits, rather than out in mid-ocean. In addition, the Japanese wish to emphasise that their buoys can function in a placid sea. And, most important, the buoys ride the tide. They are moored on a long chain so that, inshore, they can rise and fall with the tide as well as the waves. The tide is of no significance in the deep, distant oceans.

The energy captured by a Masuda Buoy can be roughly (very roughly) depicted as the difference between the movement of the waves and the movement of the buoy. Imagine a tin can bobbing up and down in the water. The air pocket in the top end would scarcely change. There would not be enough air going in and out to drive a turbine. At the opposite extreme, if the can was held fast, the movement of the waves would be at the maximum, like unconfined waves dashing up and down a harbour wall. But the pressure on the can and its mooring in a rough sea would be enormous and the can would be either smashed to bits or would break away from its anchor. These are crude, extreme examples. Somewhere in between lies the efficient use of wave energy.

The Japanese have been able, at least for the buoy, to brush aside the question of efficiency because they need only enough light for a 60-watt bulb. But if wave energy is to be used as a substantial contribution to the national grid, then the question of efficiency becomes important. And it is here that we come to the basic difference between the Japanese approach and the British. The Japanese have gone storming ahead, just as they did during the last war but this time with a better motive, in what was once thought of as the British style: the pragmatic approach.

They have used the waves for lighthouses, ocean survey instruments and observation towers and are now working on a 500-ton ship, the *Kaimei,* which should produce between 1.3 and 2.2 MW, about 100 times as much as our most ambitious schemes. It is backed by a seven-year programme and $39.5 million (about £22 million). The British Government has allocated for four devices first £2.5 million in April 1977, followed by £2.9 million in June 1978 — "drops in the ocean", as one commentator called it. The programme is subject to annual review, so long-term planning is impossible.

The *Kaimei* is 80 m in length and 12 m in breadth. It has 22 holes chopped in its hull. These lead to 22 air pump rooms. There will be one air turbine over each pair of rooms, an indication that the Japanese have not bothered with the more sophisticated designs of turbine. It is possible, by use of a device invented by a team at Queen's University, Belfast, to use a Wells turbine, something like an aeroplane propeller producing no forward thrust. In this way, the turbine can be driven by the air whether it is being sucked in or pushed out. We may well do better than the Japanese in the course of time, but we shall have lost the lead.

In what sense "better"? For Britain, it has come to be accepted that "efficiency" is supreme. The CEGB, the National Engineering Laboratory (which is developing and improving the Japanese device), the Department of Energy, the nuclear scientists — all are obsessed with efficiency. It is used as an engineering term and can be and is quantified. It is a development of what "efficiency" means to the layman. To the engineer, it implies that a device costs so much to build, so much to maintain, service and replace when bits go wrong, so much to run taking into account the cost of manpower and spare parts and loss in generation and transmission; and that it produces in wave energy so many kilowatts from so many metres of device. The way to work out the sum is to feed all the information into that famous British invention, the computer. The answer to the sum is supposed to represent wisdom. The only item left out is commonsense.

With the Masuda Buoy, efficiency is immaterial. It is actually necessary to reduce it to prevent the batteries from becoming over-charged. Where production for the national grid is concerned, the question of efficiency is obviously more important. We want as much power as we can obtain from the available equipment. But there is one sense in which we need to grasp an entirely new concept. The amount of energy available around our coasts is, like the amount of energy available to each individual Masuda

Buoy, greater than we are ever likely to need. Even if the 120 GW for a "front line" of generators was reduced to, at the most pessimistic estimate, 12 GW, we could build parallel lines of devices at 160 km or 80 km spaces. For the first time since the industrial revolution, we can contemplate a situation in which we do have elbow room.

The initial cost of construction can be and, in my opinion, should be regarded as an element in a public works programme, providing useful employment in a slump. After that the cost of maintenance, compared with the cost of running and maintaining a coal mine, will be small.

The advantages and disadvantages of the different devices are being discussed at many levels. It is a fact that the OWC and the HRS Rectifier need more concrete, while the Duck and the Raft would need more steel, which is more expensive. One leading expert has also made the point that the OWC would be able to respond to the rate of irregularities of the sea more easily than the Duck or the Raft. He put it this way: "The Raft and the Duck have the bad thing that they transmit with high torque and low speed. But wave energy can rotate an air turbine at 1,200 rev/min. So with the Duck and the Raft, you have a slow movement at first. The device has to accommodate its speed to the particular type of waves which are hitting it. If something is slow to respond, it won't be able to react quickly. And the higher the torque and the slower the speed, the bigger the machine and therefore the more expensive it is. It would be easier to smooth out the erratic output of an OWC."

Now let us see how Britain is developing the same idea – the idea of converting wave power into pneumatic power into mechanical power.

The NEL is hoping to launch, in the summer of 1979, a 1/10th scale model of its more-efficient device. The model will be 12 m in length and 3.5 m in breadth, long enough for a man to walk from one end to the other in 14 paces; slightly superior to a prison cell. The full-scale prototype would be larger than the Japanese: 120 m x 35 m. But there is no sign that we are even thinking of trying that, yet.

The Japanese have embarked on a programme covering the years 1974–1983, with a decision made in 1976, after only two years' study, to go ahead to a full-scale prototype. Their intention is to moor the ship in 40 m of water, about 2.5 km from shore, and start test generation in the autumn of 1978. Our intention is to start generation from a one-tenth scale model near Ardrossan, in the mouth of the Clyde, nine months later.

I do not doubt that our model will be more efficient. But does this really matter when both countries are pioneering the route into a new technology? Remember, too, that the White Paper policy of the Department of Energy is that by 1981 or 1982, a single device will be "identified" and resources will be concentrated on it. The unchosen people can than knock off and find themselves something different to do.

Meanwhile, the Japanese are jumping in with both feet; their philosophy, to mix one's metaphors, appears to be to push out the boat and see who salutes. We are standing on the beach, dipping our toes in the water. Soon, we shall hears cries of dismay as we watch the Japanese exporting their devices or their patents to enable other countries to use or manufacture the "inefficient" versions of their generators, while we study print-outs from computers to find something better. But later.

These views will surely be regarded as heretical by the brilliant group of engineers from whom I first learned about wave energy, the experts of the National Engineering Laboratory in East Kilbride, near Glasgow. They are the ones who produced the first major feasibility study into the subject in February, 1975. It remains, for me, the basic source of information and is certainly the most comprehensive study that has been produced; it is three years old which, in this field, makes it almost archaic, yet no-one has written anything better. This gives a measure of the talent available at the NEL, the most elaborate centre of engineering technology in Britain. It was set up by the Department of Industry and today, with a staff of 850, and 67 acres of laboratories, can cope with many of the problems facing industry. It is a dream world for anyone interested in engineering, with everything available from the biggest computers to anechoic chambers to study noise pollution. The service is available to private and nationalised industry and any employer can ask for its help. It will send off a team of experts to diagnose, for instance, gremlins in new machinery that refuses to bed down. Stethoscopes will be applied to the machine, plugged into a telephone and hooked up to the computer, which has capabilities beyond the resources of almost any private company.

The service is less well-used than it should be, partly because middle management resents the idea that an outsider might know more about how to solve its problems than it knows itself. So the senior people at NEL are not surprised when they meet resentment. Their attitude to their work is the refreshing one that one encounters when highly-qualified engineers are given the opportunity of working on subjects that fascinate them, without the pressures that come when the profit motive is supreme.

Because the NEL enjoys this freedom, it attracts our best engineers and it also, inevitably, attracts criticism. The fact that it was given the job of investigating wave energy will not in my opinion have enamoured some people in the Department of Energy where one tends to find individuals labelled as coal, nuclear or, particularly, oil. Understandably, there is resentment at the idea that outsiders might be experts on any form of energy.

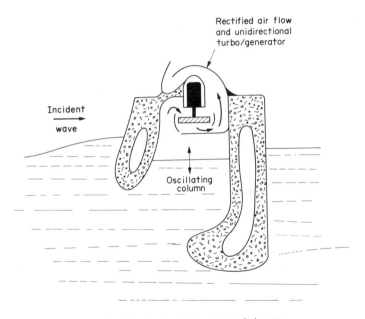

Advanced air bell of the type proposed by NEL

Figure 9.2   How the Oscillating Water Column will be tested. The shape is designed to obtain the maximum efficiency. The air bubble will turn the turbine and drive the generator with a rectified air flow which will enable the outgoing and incoming air stream to be used. In practice, the gap will be a series of holes in the side of a ship instead of the openings in the bottom of the floating power station, which the Japanese are experimenting with.

The NEL feasibility study, in February 1975, described the Masuda device as "the most promising scheme" on the grounds that it had no large moving parts, high efficiency, a valved air turbine generator system which had already been demonstrated as effective and reliable in small units, that the devices could be built using existing shipbuilding and construction technology and that it had a higher credibility rating than most of its rivals. Note the high priority given to efficiency.

Just over a year later, when the Department of Energy at last got round to distributing money for further research, the NEL was assigned the Masuda project to work on — the one it had itself chosen as the most promising. Today, the engineers there remain strong advocates of its virtues.

It has acquired a new name. It is now the Oscillating Water Column (OWC), because a column of water oscillates inside an upright concrete tube. They have rejected the name of Masuda on the grounds that they have developed and changed and improved his original idea, which is true enough; yet it is difficult to argue that he should not be given the main credit for the idea, as indeed he was in the original NEL study.

The first thing that NEL did with Masuda's idea was to test it for efficiency. George Moody, a mechanical engineer aged 31 who specialises in fluid dynamics, explained that their first step was to make one side of the column shorter than the other, to provide a gap at the section facing the most fruitful waves. This would help the waves to enter and "go up the spout". By this change — obvious with hindsight, but requiring creative engineering talent — they increased the efficiency from 30% to something like 70%. Then they built a base plate, at right angles to the longer, rear wall, and parallel to the seabed and this took efficiency up to 90%. The shape which emerges looks like an old-fashioned double-decker bus standing on its rear end, with a gap at the bottom left-hand corner for passengers (or waves) to enter. This is the sort of improvement in efficiency at which the NEL excels and at which no-one could quibble. It has been achieved, let us remember, in less than two years. It is the subsequent stages, the repeated tests and the stage-by-stage upscaling, and the feeding of information into the computer before venturing too far, which is where we differ from the Japanese.

In the NEL design, there are two pipes at the top, one for sucking in air to fill the vacuum when the water goes down, the other for pushing it out

Figure 9.3   1/50th-scale model of an Oscillating Water Column being prepared for testing at the man-made lake in Strathclyde Park. Standing alongside it is Graham Rae, George Moody's assistant at the National Engineering Laboratory, while a mechanic adjusts the rods linking the sections. The model is designed to float with roughly one-third of the windows above the plimsoll line so that engineers can see the water rising and falling. The holes in the top would normally be the channels leading to air turbines. But in the experimental models, pressure transducers and depth gauges are stationed there to give readings of the rise and fall of the water and the air pressure in the orifices.

as the waves rise. The pipes have rectifiers: that is, they rectify the air flow with valves so that it hits the air turbine from one direction only, whether it is coming in or going out. Engineers at Queen's University, Belfast, have been working independently on this aspect of the problem and have devised a turbine with fixed vanes which needs no rectifier. They compare it with an aircraft propeller, but one that produces no forward thrust.

A problem that would not be under-estimated by the NEL is that the OWC will be facing up to the full force of the waves. Its front wall will take a terrible pounding. Colin Greeves, a tough, practical engineer who has been my guide on visits to NEL, and who has a grasp of all of the activities there, put it this way: "The water particles will be going through a motion of 20–30 m diameter, and the device would try to follow it". They have to find a means of mooring it so that it is sufficiently stable to resist this torque, without being smashed by it. Here we come back to the difficult question of finding a happy medium between a fixed structure out in the sea, between five and 30 km from shore, resting on the seabed; and the loosely-moored structure which responds too freely to pressure. One has to appreciate that the movement of the structure in the waves would be out of phase with the movement of the waves inside the column, so even a fairly free, floating structure would produce energy because its movement would be at a different pace from the movement of the air bubble. Mr. R. A. Meir of the Energy Division at NEL, has done pioneering work on this aspect. He has demonstrated that a device shaped like an upturned bus could be as efficient as a fixed device, provided that the waves which it would normally generate by its own movement were cancelled out by the rolling motion of the device itself.

Mr. Moody put it this way: "If you could fix it on the seabed, you would be on to a great thing. But remember that we are talking about building the equivalent of a harbour wall in 120 m of water. So we are inclined to think that a floating device, even if it is less efficient, would actually be better. It would not be able to accept the most powerful waves but the mechanical energy could not cope, anyway. You have a finite structure so it is just as well if these devastating forces are filtered out. You are actually better off with a less efficient device. There is a sharper decline in efficiency but the output is still high."

Note that the Masuda ship will be facing the waves, with holes in its bottom, and less of the wave energy going "up the spout". The NEL ship will be moored broadside on to the waves, with the holes opening in the side and the waves hitting them with full force. So it is certain that the NEL ship will capture more of the energy and will, therefore, be more efficient. But it will be a 1/10th scale model, taking to the water probably a year after the Japanese have launched a full-scale prototype. If you were sending out salesmen to foreign countries to sell a wave energy generator, there can be no doubt whose scheme would be more attractive – the working full-scale ship, rather than the smaller, better model. And if we do decide to build the OWC full-scale, there is no way in which we could offer a prospective customer evidence of how well it works until something like, at best, three years after the Japanese had sold their patents.

Against this, the NEL have a greater problem with mooring and they are not the ones to under-estimate it as they have done a great deal of work with North Sea oil platforms. Unlike Salter and Cockerell, the NEL

Figure 9.4   The Oscillating Water Column. Inside each "ship" are three units consisting of a water column and a bubble of air. As the waves rise and fall, the air is pushed out and sucked in and the motion drives an air turbine. In the artist's impression, the turbine is enclosed behind hatches, with the hatch on the far right opened for maintenance. The arrows show the direction of the air movement as the waves outside rise and fall.

do not want the major part of their structure moving with the waves. Mr. Greeves is insistent that more work needs to be done on mooring before a prototype can be installed safely. We shall soon see how Masuda manages with his ship and it is at this point that the question of efficiency will arise once more. If, as seems likely, the ship is moored less firmly than the NEL would wish, then its efficiency will be less because more of the waves' energy will be wasted as the ship rolls backwards or rises and falls in the waves. It is then that we shall once again be faced with the question of efficiency. Masuda's ship seems likely to produce no more than 20 kW/m from its length of 80 m. Off the Hebrides, the waves produce 70 kW/m. Anything like the efficiency at which the NEL is aiming would obviously be enormously superior in output to the Masuda project.

The question, which is a political one, is whether it is better to delay in order to find the best system, or to revert to the old British habit, which the Japanese have now acquired, of a pragmatic approach.

It seems to me difficult not to feel that they are emulating the spirit of some of our early experimenters with steam energy. We produced contraptions like Watt's engine which were enormously inefficient compared with superior models which succeeded them. But, by stumbling ahead with determination, Britain did succeed 200 years ago in leading the world in moving from water power to steam power, making expensive mistakes from which other countries later benefited. Nevertheless, our pioneering efforts did give us the prime place in the industrial revolution.

It would be sad now to have to watch the Japanese charging ahead in the best British manner while we cautiously pursue a scientific approach.

CHAPTER 10

How Firm Can It be?

The most forceful argument against wave energy, and all other unconventional sources of power, is that they cannot be relied on at all times. "You cannot make an unfirm source into a firm source", as some of the experts argue. If this were true, then we would be discussing something that could never be more than a back-up. We have fluctuating demands for electricity and when there is a peak the suppliers are expected to respond rapidly. As we know, they do not always succeed — we get voltage reductions and power cuts. With wave energy as a basic supplier, the problem would be greater, the argument goes, because we could not rely on the sea to respond in the way that nuclear, coal, gas and oil power stations can do. On an unexpectedly cold day, when the seas have turned calm, what do you do? Or, more pointedly still, how do you cope when a popular programme on television, and the commercial breaks which enable viewers to put on their electric kettles, create unusually heavy demands between, say, 8 p.m. and 10 p.m. This does really happen — the CEGB can provide graphs showing just how it has to meet suddenly-varying calls.

One answer, already indicated, is that you station your generators in different seas and are never likely to be caught with all the seas calm at the same time. In addition, no-one has ever suggested that you should close down all conventional, fossil fuel power stations and rely on the waves alone. But this still leaves an important gap. If the waves are to be a basic supplier, in the way that steam has been in the past, we must have some means of building up stocks of power.

This is a problem that exists whichever method is chosen for providing our future energy. It has been, ever since electricity was brought in on a large scale, a major difficulty. The CEGB is engaged constantly in seeking ways of solving the problem. Electricity, unlike coal and gas, cannot be stocked. Numerous ways are being investigated to get round the problem. Among them are advanced batteries, storing heat at power stations, turning water into hydrogen by electrolysis and then converting it back to electricity using a fuel cell, and using compressed air. One day, any of these methods may become practical. But for the present the only available method is pumped storage.

It is controversial, particularly with environmentalists. It has, until now, meant damaging some of our scarce mountains and hilltops. No matter how much care is taken — and a great deal has been — there is no way in which a mountain can be regarded in quite the same way if there is a reservoir where once there was a lake and a power station down at the bottom. The CEGB and the North of Scotland Hydro-electric Board have done wonders in safeguarding the countryside but any development of this activity will surely cause problems.

In a paper delivered to Oceanology International, Dr. P. A. A. Back, a partner in the consulting engineers of Sir Alexander Gibb, listed 15 possible sites for further pumped storage and aroused a stir of disquiet in the areas that he had named. I shall return to his invaluable contribution to the discussion but for the moment no-one need doubt that the very people who are concerned with the environment, and therefore natural allies of benign energy sources, will be most aware of the need to safeguard the countryside.

Further objections are that it is expensive to build the high-level reservoirs that are needed, that it is wasteful because it requires four units of power to push the water uphill in order to get three units as it comes down (though this is of little concern if the waves supply the energy) and that even if these objections are overcome we do not have the area necessary, the miles of mountains and empty spaces, that other countries enjoy.

Our present capacity is for five hours' supply, if the whole country should suddenly need to use the available power. The main stores are Ffestiniog in Snowdonia, which was the first and has been, since 1963, giving 360 MW; there is Foyers on Loch Ness with 300 MW capacity; Cruachan on Loch Awe with 400 MW; and a new station being built in Dinorwic in Wales with 1,800 MW capacity. The total is 2.8 GW. There is a plan for a new scheme at Craigroyston on the eastern shore of Loch Lomond, which would have a capacity of 3,200 MW and, if local opposition can be accommodated, that would take our capacity nationally to 6 GW, which would be a considerable item but not enough to turn wave energy into a firm source. It is still only one-tenth of peak demand in winter.

Dr. Back has had the courage to identify other possible sites. He noted in his address that renewable sources of energy demanded storage and, with the limited number of inland sites available, turned his attention to the coast, using the sea as the lower reservoir with an upper reservoir on the land nearby; with an alternative of using the sea as the upper reservoir and disused mines or "specially excavated deep caverns below sea level". It is this second option which may provide a solution, acceptable to everyone.

Dr. Back began by explaining that the general pattern is for the upper reservoirs to be gradually depleted between Monday and Friday, and topped-up at the week-end. The reason that this happens is that there are generally six hours available, at night, for pumping water up to the reservoir, and then six hours when it thunders downhill to meet peak demand. As it works at 70% efficiency (roughly), the water level is gradually depleted.

Dr. Back laid down the criteria for a worth-while site. It must have a minimum plant installation of 600 MW, a minimum head of 100 m and an upper reservoir within one mile of the shore. Artificial lining might be necessary, to prevent salt water damaging the land, or sites must be chosen with low permeability. There would need to be underground shafts, tunnels and caverns – possibly 150 m long, 25 m wide and 45 m high. The stability of the land, and particularly the cliffs, must be safeguarded. There must be close attention to the biological problems associated with seaweed, barnacles and mussels on the intake/outfall and within the cooling water system. In addition, "the environmental impact of a coastal scheme is likely to be of considerable significance particularly since those areas having high coastal cliffs and therefore attractive for pumped storage are also attractive for tourists. In addition, many such areas are also notable breeding grounds for sea birds."

A formidable list. I have quoted it at length because I do not wish to be accused of being unfair and because the practical problems need to be spelt out, if an informed public is to consider the situation. But I have another purpose.

Dr. Back was discussing methods of storing power if we turn to sources of energy supply other than fossil fuels. Those "other" sources are listed by him as solar, wind, wave, tidal, geothermal — and nuclear. In all the argument about nuclear power so far, its advocates have been discreetly silent about the fact that if we increase our dependence on nuclear energy, a vast extension of storage, pumped or otherwise, will be needed, And if the people who are advocates of nuclear power have their way, then environmental considerations are likely to have a low priority indeed, while the engineers and scientists most concerned with finding benign sources are precisely the sort of people who will appreciate that the countryside is as much a part of our future as the genetic legacy we leave behind us.

Is there any alternative? Do we really have to choose between digging up the countryside for coal mines and pumped storage, whether it be for nuclear or wave energy, and going without power? It is Dr. Back who indicated the real alternative. It is underground storage.

We could, he noted, use the sea as the upper reservoir and allow the water to pour into an underground cavern. Instead of pumping the water uphill in off-peak periods, we would let gravity do the work and then use the off-peak periods to pump it back up to ground level. The energy would be obtained as the water flowed down.

It would still, of course, be "expensive" in the sense that you would need more energy to push water up than you would obtain as it fell to the bottom of the cavern. But provided that you are using renewable sources rather than fossil fuels, the cost would be of no concern.

Dr. Back identifies the factors which are needed: "a void of about four million cubic metres at a depth of 500 m, or a proportionately larger volume at a lesser depth, would accommodate a 600–800 MW installation for six hours' generation a day." He suggests that disused mines near the coast would reduce costs. He has examined 31 mines which might be useful for this purpose and notes that they are "fraught with challenging problems for the civil engineer". His conclusion is that the iron ore mines of Cumbria are the best bet. They are between Maryport and Whitehaven, in the very area of Cumbria where Windscale stands. Local advocates of increased dependence on nuclear power, who are concerned with employment in the area, might care to contemplate just how much work would be created by the development of underground pumped storage in disused mines.

But, as Dr. Back says, this is of limited value and there is no way of avoiding his conclusion that some areas of countryside would be needed above ground — unless some other method can be found of storing energy.

There are, as I have indicated, numerous ideas being examined by, among others, the CEGB, mainly in its Research Laboratory at Leatherhead. One such idea, which has attracted little attention so far, is to build a man-made "cavern". As one of its leading scientists, Dr. Anthony Hart, put it to me, it would resemble an underground car park, spiralling down below the surface with the water whirling to the bottom where it would hit a turbine. The disturbance to the surface would be no more than a big plughole, with all the plant buried from view. It would cost money, of course. So do nuclear plants and coal mines.

I have given this brief survey of the storage problem because it is necessary to answer the telling argument that wave energy could not be regarded as a firm source. That view is challenged by some of the best brains examining wave energy.

And we must not allow the CEGB to get away with the argument, which is almost gospel among its top officials, that nuclear energy is wonderful because you can bring it in by pressing a button and that other sources are unreliable. Both arguments are untrue.

Addendum: The Bathtub Syndrome

A group of wise men sat on a hillside in Sumeria in the year 3500 B.C. and scratched out designs on clay tablets. They had to decide whether to go from research to development of a new invention called the wheel. They had to consider its cost and its efficiency and they brought a proper sense of scientific method to their discussion. "After all" they said "what do we need it for? We can all get about by walking. And remember, walking is *safe and clean and reliable.*"

The last sentence contains exactly the words that were spoken to me on the subject of nuclear energy by one of our most eminent scientists. We were having a vigorous exchange of views during a break in the Wave Energy conference held at the Heathrow Hotel in London in November, 1978, A.D. The speaker was one of the more enlightened scientists employed by the CEGB. He was a key figure in early studies made by the Board into wave energy. But he had never deserted his first love, nuclear power, in which he had been educated and to which he had devoted many years of his student and wage labour. He believed implicitly that nuclear power stations were worthy monuments to 20th-century science.

"What", said another distinguished scientist, "are you going to do with a wave power station when it shows metal fatigue?" I replied: "We shall encase it in a glass block, because everyone knows that glass is unbreakable, and bury it in the sea alongside the vitrified Windscale chimney which is still sitting, irradiated, over the Cumbrian countryside while nuclear engineers decide how to dispose of it."

The conversations that I have reported took place when wave energy sank into a trough. It had scored an own-goal by asking a distinguished firm of consulting engineers, Rendel Palmer and Tritton, to estimate the cost of wave-electricity if you built one wave energy power station from each of the main devices with 2 GW installed capacity. The only sensible

103

answer to such a question at that time would have been: Don't be bloody silly. What answer would James Watt have given if, in 1800, when steam was producing 11 kW, he had been asked a similar question? Remember that Ernest Rutherford died in 1937 convinced that nuclear energy would never be economic. The Wright Brothers calculated that flying would cost $30,000 a mile and they would have been right if costs were based on building, say, one Jumbo jet.

An appealing part of wave energy is that it is modular. You can build one unit which produces electricity immediately, and repeat or improve it almost indefinitely. Unlike railways, which are committed to one gauge, there is no restriction on placing one unit alongside another, with the lessons learned from the pioneer applied to later generations. You improve as you learn from experience and, as with motor car engines, the cost comes down in proportion to the amount of serial production. It is totally different from building one coal mine or one power station or one tidal barrage, all of which require enormous capital investment in a once-off operation. The range of wave power goes from the navigation buoys which light up one 60 watt bulb and drive a flasher unit from a 12 volt battery to the 2 MW power station, the *Kaimei*, built by the Japanese, to the fleet of units which would add up to a full-size power station rated at 2,000 MW. The question might never have been put to RPT if it had been realised that the conclusions would, inevitably, be misinterpreted by the public. An extraordinary feature of the Heathrow conference was the contrast in response to the findings among the engineers and scientists on the one hand, and the Press, among whom I sat.

The reporters started to leave after they had heard Mr. Peter Clark, the chief of the RPT team, state that wave electricity from the OWC, the Raft and the Duck would cost between 20p and 40p/kWh, from the HRS Rectifier 30p–60p and from the newer Lancaster Airbag perhaps 5p–10p, compared with 2p an hour which is the notional figure used by the CEGB for conventional and nuclear electricity. The Airbag, which was the one device that seemed to be financially serious, was played down in other ways. Dr. Freddie Clarke said: "Our knowledge of this device is at a relatively early stage." Professor Michael French, its inventor, himself said: "Questions of bag design and durability constitute the most critical area of uncertainty with this device and may possibly make it impractical." Mr. Grove-Palmer said: "The feasibility has not yet been established. We shall be subjecting it to tests soon." The impression was that its comparative cheapness was due to its newness and the absence, at that time, of an opportunity to subject it to extensive testing.

But among the leading figures engaged in the research programme, there was no dismay. They understood that the report was conditioned by its terms of reference. They did not appreciate how the public would see it. Stephen Salter appealed to me not to criticise the report. He, and the other experts, were delighted with the amount of research that had gone into it and regarded it as a valuable document. The figures for cost were understood by those closest to the scene and were in no way a cause for concern.

The major objection to the cost-estimates is that too much was unknown for any sensible answer to be given to the question. When, after all, has anyone seen a "costing" from a consulting engineer with a range of 100%? What would be the reaction of a prospective house-buyer if told that a new house would cost between £30,000 and £60,000? Surely he would turn on the surveyor and ask just which figure was right, as it would make quite a difference to the mortgage repayments.

The conference was attended by Sir Hermann Bondi, Chief Scientific Adviser to the Department of Energy, who would not be accused by anyone of timidity of thought. I had had the opportunity of meeting him to discuss wave energy and I had put it to him that the British disease nowadays was feeding everything into computers and committees. He replied cheerfully that I sounded like one of his own memos. He said, however, that he had never been under pressure to stop spending money on wave energy. "The only pressure has been whether I could spend more usefully. I myself have been concerned about what we can do. A reckless going-ahead into the wildest waters in the world, with not fully-tried and developed devices, is something that I don't want to do. Our policy is to support, encourage, advise and fund all that we see on the market at the rate it will take."

Sir Hermann is less concerned with the numbers game than some of the CEGB people, who are obsessed with questions of cost and undismayed by the fact that their essays in futurology have become a joke. He is extremely aware that cheapness is not all. As he put it: "We face an uncertain energy future. One must spread one's risks by insurance. People must have plenty of options available. They have a right to choose. If they don't choose the cheapest, that's their right. If people were to say that they didn't want nuclear power, I would want to tell them what it would cost them. People don't always buy the cheapest motor cars. I would like to get them to the same point in energy. It must be a matter of choice. The last thing that I want to do is to ram anything down anyone's throat.

"Some of the arguments of the anti-nuclear lobby are funny. There are also some very serious considerations, proliferation in particular. Our policy must involve nuclear and fossil and renewable elements. I want to have them available and costed, in case other forms become unacceptable or unavailable. I regard 'unacceptable' as being as important as 'unavailable'. If it costs x times as much, it may be that one day people will want to pay that premium. So I am the last man in this field to throw anything away."

Then he added: "When people talk about cost, I say to them that if they had to choose between their motor cars and having electricity in their homes, they would abandon their cars. But they spend more on their cars than they do on electricity."

Faced with such reasonableness, presented by Sir Hermann with charm, how can anyone demur? One must return to the Heathrow Hotel to discover the reality. The question of cost was dominant, but so was a tendency that was pointed out to me by one of the leading engineers in wave energy. He is a key figure in work on one of the foremost devices. He said to me: "ACORD (Advisory Committee on Research and Development) and WESC (the Wave Energy Steering Committee) are staffed by people attached to research, which is very different from development. I sometimes wonder if they will ever cross the line." Because we usually couple R & D, we tend to think of them as natural partners. They are not always so. The present stage has been described by some as "the bathtub syndrome". Engineers as well as scientists will explain that in a wave tank in a laboratory they can test a design, modify it, and then feed in an identical sea. When they go into the real water, they may have to wait weeks or months for comparable conditions. Everyone understands that. But in the end, it is the real sea which will have to be tackled.

The official report of the conference proceedings echoes this feeling:

David Ross: Isn't it becoming rather evident that in Britain we are seeing the alternatives of efficiency and cost-effectiveness as the only realistic things and that in the end, possibly by the early decades of the next century, we shall have a very high percentage of zero while the Japanese are going ahead now and producing something which is being described as reliable rather than efficient, but is producing energy?

Professor D. R. Swift-Hook of the CEGB: The suggestion that David Ross seems to have made is that there is a difference between efficiency and cost-effectiveness when your energy comes for free. I am afraid that just doesn't hold water, if that's the right phrase in this context. . . . High efficiency is essential for good cost benefits.

In such an atmosphere, it was difficult to recall that we live in a world calmly using up its finite supplies of fossil fuels, a world in which millions of people will live without ever switching on an electric light and without ever enjoying any of the benefits that are available from the intelligent uses of energy.

I could not help being reminded of the Dyson cartoon showing Clemenceau leaving the Palace of Versailles, saying to Lloyd George: "Curious, I seem to hear a child weeping." We had met not in marble halls but on plastic floors. If a child had wept, we would not have heard because the Heathrow Hotel, wedged between the runway at London Airport and the Bath Road, is insulated against noise and the air is, of course, conditioned.

***

That was a low point for wave energy. But worse was to come. One year later, in November, 1979, the Department of Energy produced Energy Paper 42 on wave power. I was given an advance copy by a friend in the Department and I howled in agony as I opened it. On the first page, in a Preface by Dr. Freddie Clarke, was a repetition of the figures of 20p–50p/kWh compared with around 2p for "normal" electricity. A footnote mentioned that this represented the situation at the end of 1978 – and even this was an under-statement as the RPT report had been written in the summer of 1978 and was by then nearly 18 months out of date. But it was obvious that commentators who were not specialising in wave power would not appreciate the significance of the footnote and that once again great damage would be done.

Friends in the Department and at Harwell appreciated the point and a great deal of work was done to ensure that some, at least, of the reviewers would be up to date. But inevitably damage was caused. It was, as an

18th-century Frenchman remarked, worse than a crime; it was a blunder. One could not help but contrast it with the smooth way in which the nuclear lobby operates, making even Three Mile Island sound like a success story. But the good, guileless people concerned with alternative technology are not organised in that way.

The real pity was not in the unfortunate impression given by this reference to outdated costs but in the fact that the rest of the publication was largely ignored. It is a superb document which could and should be studied with appreciation by everyone concerned with our energy futures and with the way in which Government scientists are accepting and tackling their responsibilities.

Much of the information is highly technical and had never been available before. For general readers, particularly those concerned with the effects on the environment of a line of 1,000 km of floating power stations, there is an invaluable chapter recording the investigations of a Technical Advisory Group containing experts from the Nature Conservancy Council, the Scottish Marine Biological Association, the Department of Agriculture, Fisheries and Food, and the Highlands and Islands Development Board. Their conclusion is that "no environmental effect has emerged which would raise major doubts".

They have investigated the impact on the beaches and conclude that there might be less erosion from winter storms, which is if anything an advantage. There is a serious discussion of the effects on herring, salmon, cod, haddock and lobsters. They remark that a wave converter fixed to the seabed — such as the HRS Rectifier and the new design for an Oscillating Water Column (which is explained later) — would increase the hazards to the lobster fishing boats and pots and provide "a benefit to the lobster". Alas, poor lobster, this type may not be chosen for the Outer Hebrides. (Information received from Japan about the floating Masuda ship, the *Kaimei*, indicates that the fishermen may also gain a bonus as fish collect around it, as they do around all man-made structures in the sea, and increase the catch. But the catch here is Catch 22 because one is in danger of being accused either of interfering with fishing or with increasing the speed at which we deplete the ocean's stock of food. The realistic answer is that we should insist that the size of meshes is properly controlled; and that the task of the fishermen will be made easier by the wave power stations. In that way, the fish *and* the fishermen should have an easier time.)

The care with which these questions have been considered can be gathered from the following passages: "Perhaps the most important factor to emerge so far is that, since the herring spawn on the gravel areas of the seabed in this locality (around the Outer Hebrides), major disturbance or removal of the gravel for construction purposes would appear to be inadvisable. . . . Salmon migrate over long distances. It is possible to infer that the area immediately to the west of the Outer Hebrides may intersect some of the routes. In the locations concerned, the salmon probably swim near the surface and the problem arises as to whether a line of wave energy devices could deflect them off course or whether the fish would simply swim through the gaps. Moreover, the converters might create an environment favourable to large colonies of predatory birds which would feed on the smolts, and of seals which might feed on the adult fish. The rapidly-expanding colony of seals on the Monach Isles might recognise the converters as an advantageous additional habitat. . . . Further studies are planned to examine the nature conservation implications. A major installation of wave energy converters several hundred kilometres long could not take place in a short time but would be spread over many years. Experience gained with the early modules would be available in good time to take corrective action if that proved to be necessary."

What about ships? Wave energy converters will certainly be, as the Energy Paper puts it, "a hazard to shipping", particularly as they will have a low freeboard which will be difficult to see even with radar. An easy solution to that is to put a light on the top of a pole; there will be no shortage of electricity to fuel it! Ships do have to try to avoid many obstructions at sea, including the navigation buoys which warn them of sandbanks and wrecks, oil platforms, weather ships and piers. Gaps would have to be left between the power stations and it is suggested that they should be 2 km wide. "No new matters of principle appear to be involved."

There is one other hazard which has not been met before. Part of the wave spectrum will be reflected and this will "produce a complex pattern in which there will be increased wave heights, standing waves and an increased incidence of breaking waves". In other words, a man-made storm sea. "However, it is possible that this effect will be confined to a distance close to the converters and the declaration of an exclusion area around the converters may be sufficient to cover this point. Moreover, there would be a counter-balancing advantage from the fisheries point of view of calmer water behind the converters. . . . Past experience has shown that an exclusion area would not entirely eliminate the consequences of the hazard or the chance of a collision, since human nature ensures that commercial attractions (prospects of higher fish catches) sometimes override prudence.

"The treatment as a statistical problem of the incidence of collisions between ships and fixed installations is receiving attention in the context of the North Sea and no new work appears to be needed specifically for wave energy at present."

The extracts themselves show the care with which the problems are being considered. But they also reveal the leisurely pace of advance. Would we ever have drilled a hole in the North Seabed, or indeed ever have built a harbour or a pier, if every conceivable consequence had had to be analysed first? The Energy Paper does seem to be visualising a sea alive with cargo ships, cruise liners, fishing boats, rowing boats, drunken sailors, short-sighted skippers and armies of trippers lining up for a pleasure trip round the wave energy station. This is not exactly like the sea off the Outer Hebrides. Practically no-one uses it.

There is one environmental disadvantage listed: "One of the most important aspects from the visual amenity point of view will be the method and route by which the energy is carried to the mainland. . . . A system feeding electricity into the national grid is unlikely to be able to avoid transmission across Skye. Very detailed and careful transmission route planning will be needed." The best solution to that problem would be to put the cables underground. It will cost a lot more, but nothing like the cost of, for instance, the fast-breeder reactor which the Highland Regional Council is asking the Government to build at Caithness for an estimated £1,500 million.

What would wave energy mean for employment prospects in the Outer Hebrides? It would, says the energy paper, "present many new opportunities for the development of both the traditional seafaring and new industries as a substantial contribution to local economic wellbeing. . . . The Western Isles Island Council has stated that its primary objective is the reversal of the trend of population decline due to the persistent selective out-migration from the area, and the improvement in the situation of the population in terms of employment. Installation of wave energy could assist in two ways: the labour force needed to operate and maintain the system; the possibility of utilising some of the power to establish new industries in the area."

The technical content of the document is superlative. Nothing comparable exists anywhere. Much of it bears the hallmark of Clive Grove-Palmer whose influence on the wave energy programme has undoubtedly been a major national contribution. He has the gift of mastering the most complicated engineering and scientific concepts and then presenting them in a way that any reasonably-informed reader can understand. His background is a degree in structural engineering and electrical instrumentation. During the war he was a scientist on minesweepers, devising ways of coping with magnetic mines. Then he worked in naval construction, instrumentation for heat transfer systems and disposal of nuclear waste. This combination made him the ideal choice for wave energy and the people at Harwell who realised this and chose him to head the Wave Energy Steering Committee merit great respect.

They also deserve congratulation for keeping him in his key position during the years when it seemed that the whole programme might be in danger and Clive pressed ahead, always available as a balanced, sustaining voice. He has exactly the qualities of a naval ship's captain who refuses to be flapped when the weather turns nasty, the boiler blows up, the rum runs out, the main guns are jammed, enemy battleships and submarines are gathering, a magnetic mine is clinging to the stern and the chaps with dry feet back in the Admiralty are issuing idiotic instructions.

The change of Government in May, 1979, had no immediate effect on the wave energy programme. There was no announcement of a further grant of funds that Spring but, given the political atmosphere, that may have been an advantage. At all events, money continued to reach the research centres. And then on June 13, 1980, a decision was taken secretly. It was to stop further funding for the present of the two most-advanced British inventions, the Raft which was ready to be built as a full-scale prototype producing electricity, and the Duck which still needed some work to be ready to take to sea as a finished unit but did have a solution to what its critics had claimed was its major problem, the power take-off, and was ready for deep sea testing of parts of its mechanism. Instead, funds were to be concentrated on developing the Japanese invention, the Oscillating Water Column, to which Britain had made only marginal contributions, and on the three newest devices, the Lancaster Airbag, the Lanchester Clam and the Bristol Cylinder.

All three were between two and three years behind the leading devices and the policy meant that the Government was putting off the really big and hard decisions: instead of investing major amounts in floating test-beds, work was to continue on 1/150th-scale toy boats in laboratory tanks, the bathtub syndrome once more. Most insultingly, Mr. Salter was told to concentrate on "generic spine research", the spine to be used on someone else's invention! The best answer to that one is the title of Chapter 8.

Sir Christopher Cockerell took the news philosophically: he had had similar rebuffs when he was trying to win support for the Hovercraft.

The amount of money available, taking account of inflation, was reduced by about 30%. The Government behaved shiftily over its action. A month after the decision, on July 21, Mr. Frank Hooley, MP, asked about our wave energy programme in the Commons and Mr. John Moore, Energy Under-Secretary, carefully avoided revealing the details. Indeed, although word circulated, the inventors themselves were not told officially until September 25, the day before Mr. Moore was to attend an official "opening" for a new testing tank at the Cockerell headquarters. At the ceremony, Mr. Moore spelt out details of new grants (for instance, £125,101 for Lanchester Poly — right down to the odd £1) without mentioning the cut-off for the Duck and the Raft.

People close to the scene were convinced that the CEGB had played a major role in encouraging a cutback. Plainly, there is little hope of a major investment in benign energy sources if we are to go ahead with expenditure of thousands of millions on nuclear power. But the timid approach had been pioneered by the previous Labour Government.

How was it that Mr. Benn, not normally a Minister lacking in imagination and dash, was persuaded to move so sluggishly? Since leaving office, he has spoken of the way in which the civil service mandarins obstructed him. My view is that a stronger Minister would have stood up to the pressure better. The evidence for that is to hand.

Mr. Benn received me on September 14, 1978, after the first edition of the book had been written, but six months before it was published. We had an agreeable exchange of views and at the end of it he told me that he had no wish to keep secret his briefing and he handed me the documents that he had been using. They contained photostats of an article that I had written in the *New Statesman,* and of a correspondence that had followed it, plus the civil service comments on it. Only the name of the author had been removed.

The briefing said:

> "The Department gave generous background briefing to Ross but did not have a preview of the article. The article argues, in emotional terms, that the UK is not really trying to develop wave power rapidly whereas the Japanese are successfully doing so. He further argues that our programme is poorly motivated because of official and ministerial commitment to nuclear and coal."

That is an accurate summary of my attitude and I admire the unknown author for putting it so succinctly.

> "The rate of growth of spend can be increased by accepting a greater risk of financial waste. Corners can be cut and programmes run in parallel, etc. The risks are justified in war-time situations but few would judge that the wave energy programme should be seen in such desperate terms. . . . The Japanese have pushed ahead rapidly with a single device and will be carrying out large-scale sea trials shortly. By contrast, the UK has a more broadly-based, scientific programme. It remains to be seen which approach succeeds but three points must be borne in mind.
>
> "(1) We have a stake in the Japanese work by virtue of our £300,000 contribution to the joint I.E.A. programme (they have no such stake in ours).
>
> "(2) Japanese industry has offered very low prices for plant because of low order books and this has contributed to their rapid transition to large-scale trials.
>
> "(3) The Japanese had many years' start on the UK because of Masuda's work on light buoys.
>
> "We have no evidence that anyone in wave energy work feels constrained by shortage of funds – which is not to say that they are all happy. Some are looking for long-term financial commitment by Government to a wave energy programme. This is being considered. . . . Wave power at any cost is an objective that could very easily be set and could equally easily be achieved. It would not, however, serve the economic interests either of this country or even of wave power. David Ross appears quite unable to appreciate this fundamental point."

This is a jolly good summary of the case against all of us urging greater commitment to alternative technology and I should like to congratulate Freddie, or Don, or Gordon or whoever wrote it. They are all conscientious people. They were writing within their brief. But what about the Minister? He approved or accepted the terms.

Why did he not turn around to his mandarins and point out that, for instance, it was not only Japanese industry which had low order books; our construction, steel and shipbuilding industries were also in the doldrums, so why could we not encourage them to "offer very low prices for plant"? Note the emphasis on our "scientific" approach, as opposed to "sea trials".

I accept the comment that not everyone engaged in wave energy research is "happy". Happiness is not an absolute but comparative.

Long-term financial commitment was "being considered" and apparently still is, more than two years later.

Mr. Benn's wall was decorated with a National Union of Mineworkers' banner. It would have been better if he had also displayed a reproduction of Hokusai's *The Hollow of the Deep Sea Wave*. It appears on the wall of the office of Laurence Draper, Britain's leading expert on the waves, at the Institute of Oceanographic Sciences. He accepts it as an imaginative depiction of the nature of a wave, as confirmed by high-speed photography in the 20th-century. It was a case of art anticipating science by 300 years. Mr. Benn is a case of politics not anticipating engineering.

Mr. Benn did good work in resisting the nuclear lobby and he does deserve the credit for having presided over the first cautious investment by the Government in alternative technology. He missed an opportunity of going down in history as the politician who dared to over-ride the caution of the civil servants and the prejudices of the nuclear scientists who surrounded him. But it was his fault, not theirs.

***

Meanwhile, work on capturing wave energy went ahead – with some major engineering advances. Costs came down, problems that appeared almost insoluble were overcome and a second generation of devices emerged. By the winter of 1979, four devices had been rated on the RPT scale as producing electricity for below 10p a unit. Of the "first generation" (the Duck, the Raft, the OWC and the HRS Rectifier), only one was included: the Duck, which was down to 4p–5p per kWh. The other three were all "second generation" inventions: The Lancaster Airbag, the Bristol Cylinder, the Lanchester Polytechnic Clam. It should be emphasised that not too much significance should be put on these cost-ratings, which change rapidly. This is a brief list of the major developments:

Salter's Duck: The great problem with it, said its critics, was the power take-off (Fig. A.1). The Duck would nod slowly and this movement had to

Figure A.1.   Salter's hand-made model of the power take-off. There are gyroscopes at each side of a tacho-generator.

be converted into the high angular velocity needed to produce electricity. It was doubtful if it could be done at all, and if a method were found of converting the slow motion into fast motion, by gears, levers and chains, then it would be too complex to build or to last for long. Mr. Salter recognised the force of the latter point. He has admitted that no seal had been devised that could keep out the water indefinitely. The motion had to be taken from the Duck to the spine and sooner or later the sea would get in and corrode the workings.

Mr. Salter was, as he put it, "getting nowhere" and "beating my head against a wall trying to think of a way to make very low speed electrical generators." He decided to discuss the problem with Professor Eric Laithwaite of the Department of Heavy Electrical Engineering at Imperial College, London. Mr. Salter told the Heathrow Hotel conference: "Professor Laithwaite said immediately that it was impossible. But then he suggested a particularly intriguing way of getting power off a Duck.

> "Inside the Duck we build two large gyros arranged to spin in opposite directions. Their axes are perpendicular to the Duck's axis of nod. I am sure that you will all instantly know exactly what will happen to the gyro gimbals when the Duck nods. . . . You will also remember the equations relating output torque to spin speed, disc inertia, and input angular velocity and will have no difficulty in working out the way to load pumps acting between the gyro Duck frame and the Duck casing so that the Duck feels the right damping coefficient. . . "

This statement was, I am certain, intended as a good-natured nudge. But it is an excellent example of the way in which Mr. Salter does not realise how sharp his elbow is. One could sense many of the distinguished audience bristling. The gyroscope, that neglected invention of the French scientist, Foucault, is the sort of item that many students tend to put on one side soon after starting their university courses. Mr. Salter had irritated his listeners and their rather superior comments could be heard as they chatted after the session. I heard some of them; so did Mr. Salter. But it is interesting to note that not one delegate challenged him in open combat in the formal question-and-answer period. Those who might deplore his manner will just have to accept that genius can be difficult, but is more stimulating than people with smoother graces.

Mr. Salter had gone to see Professor Laithwaite in search of a solution to one problem — to convert a slow movement into electricity. But later, as Mr. Salter thought about it, he realised that he had within reach the answers to not one but three of his difficulties. In addition to solving the problem of power take-off, the gyroscope would provide a new "frame of reference". He had until then needed the spine to remain relatively stationary. It would stay mainly upright while the individual Ducks nodded at different moments and evened out the tendency of the waves to set the spine spinning. But with the gyroscopes inside the Duck, they would serve as an independent entity, and there would be no need to transmit the nodding motion from the Ducks to the spine. This would also reduce the torque on the spine.

The third advantage was that the gyroscope could serve as a flywheel. At that moment, as Professor Laithwaite has explained to me, it ceases to be a gyroscope because it is no longer free to turn about a different axis. But it does become a power converter. It drives a fluid into a hydraulic motor, which in turn drives an alternator. Or it can be diverted to another hydraulic motor/generator, which makes the flywheel go faster, providing the flywheel storage which is needed.

Perhaps a crude comparison can be drawn between the way in which an alternator charges a car battery when the demand is low; when the engine is idling or switched off (or when the fan belt is too slack!), and the lights, windscreen wipers and booster fan are all on, the battery feeds back some of the power it has accumulated. As always, it is Mr. Grove-Palmer who sums up the process most succinctly: "The nodding motion gives you power out *or* an increase in the speed of the gyro." Mr. Salter has put it in greater detail (in an address to an energy conference at Gothenburg in October, 1979):

> The hydraulic fluid, almost certainly oil, "will drive a swash plate motor at each end of a gyro shaft in parallel with another driving an electric generator. If the pressure tends to rise as a result of a burst of wave energy, the angular deflection of the swash plates will increase so as to allow the extra energy to speed up the gyro flywheel. If the flow of energy is at exactly its mean value, the gyro swash plate motors will move to their zero displacement angle and all the oil will flow to the motors driving the electrical generator. If, during a lull, there is less oil from the ring cam pump than is necessary for the generator drive, then the swash plates on the gyro motors will swing over so that they pump and draw the energy deficit from the gyro disc."

Some other quotations from Mr. Salter's account (at the Heathrow Hotel, at Gothenburg, and at a London conference of the Institution of Electrical Engineers on January 30, 1979) help to illuminate just what he has achieved:

"We can use the gyro spin for energy storage. At full speed each gyro stores about three-quarters of a megawatt hour so that very little speed variation can provide the few minutes of storage necessary for wave power. We have a completely opaque barrier between ourselves and the salt water. The gyros do not know that they are at sea."

(Heathrow Hotel)

"The gyros can store prodigious amounts of energy, enough to provide a completely smooth output. The energy stored in each gyro spin is about the same as the kinetic energy of a 747 airliner in flight and is enough to run a megawatt for three-quarters of an hour. . . . If any readers think that this proposal is far-fetched, I draw their attention to the method of ship stabilisation developed by Sperry and used successfully in the Italian liner Conte di Savoia of 41,000 tons. Sperry used three gyros of 100 tons weight spinning at the high speed which we are suggesting and solved the high bearing load with pre-war technology."

(I.E.E. conference)

"The reserves of energy instantly available from the flywheels make for better control characteristics than any land-based system, whether steam or hydro. The entire Duck string constitutes a spinning reserve capable of stabilising the grid rather than causing it problems. Whenever a pumped storage scheme is proposed, the generating boards emphasise the value of spinning reserve which is said to be worth hundreds of millions of pounds a year, even if never used. The Ducks are claiming those millions for being a short-term but instantaneously responsive storage scheme. They will leave pumped storage to do its proper job of overnight working."

"The vast amount of flywheel storage means that every piece of electrical equipment from the shaft of the generator to the land connector can now be rated at its mean rather than its peak rating. We can deliver some of the electricity when the consumer wants it, rather than when the waves provide it."

(Gothenburg)

His plan is to use four gyros of 17 tons — one-sixth the size of those used for an ocean liner.

How, exactly, does a gyroscope function? A scientist will usually start to explain by pointing an index finger upwards and a thumb at right angles to it, and try to make revolving movements. Professor Laithwaite uses a bicycle wheel. A distinguished mathematician says that whenever he tries to describe it, he is driven eventually to tell his students to go back to equations. Mr. Malcolm Cloke, who is on the staff of ETSU, has produced a brilliantly compressed explanation and has devoted great patience to explaining it to me. He also refers inquirers to the *Encyclopaedia Britannica,* which has a four-page explanation. Mr. Salter lamented at Gothenburg: "Our only problem has been that physical arguments and mathematics are not sufficient to persuade critics. The feel of a working model is necessary." So Mr. Salter ("he is a superb craftsman", as one of his admirers noted), got to work in his laboratory and produced a beauti- fully-engineered model of the power take-off. He took it to a conference of NATTA (the Network for Alternative Technology and Technology Assessment) at Milton Keynes in August 1979, and said most flatteringly that he had brought it specially to show me. It is 75 cm long and plugs into the national grid. You can hold it and try to turn it and feel the gyroscopes inside resisting your pressure.

The gyroscope is used nowadays on ships to activate stabiliser fins. The gyros control the fins, which would otherwise lack guidance. As Sperry, the manufacturers put it: "The ship rolls. The gyroscope tends to remain stationary." You can perhaps most simply visualise it as a wall in space, with a stream of hydraulic fluid coming at it from different directions and being bounced back into a pump.

The key word is precession. It means that the gyro cage reacts against being moved, so long as the disc is spinning. High pressure oil is driven into the generator and any surplus goes into the flywheel motor, where it is stored. It will give, in Mr. Salter's view, a 45-minute warning to the CEGB when the sea is turning calm, far longer than the CEGB enjoys when, for instance, a much-advertised television programme disappoints the viewers, who move in their millions to the kitchen and switch on electric kettles and lights. The CEGB is good at responding to a sudden peak in demand — we don't often have power cuts these days — and the Ducks, "a varying and partly unpredictable supply of electricity", to quote the original official view, will actually help to lessen the very problem which they were said to represent.

It was at the Gothenburg conference that Mr. Salter suddenly rounded on his critics. He did it in a style that Swift would have appreciated. He noted that Mr. Glendenning had said that a reliable system "is one which is simple" and that Dr. Bellamy had said that "simplicity should be the over-riding design principle". Then Mr. Salter said:

"There are no recognised units for the measurement of simplicity and the various national standards institutions have not suggested approved levels for the engineering profession. But we can turn to the *Oxford English Dictionary* for a clue. There are 12 columns devoted to the concept. They begin harmlessly enough with:
" 'Free from duplicity, dissimulation or guile, innocent and harmless, undesigning, honest, open, straightforward (so far so good) . . . Small,

insignificant, slight, of little account or value, also weak or feeble (this is all a bit much but we have only covered two columns so far) . . . Deficient in knowledge or learning, characterised by a certain lack of acuteness or quick apprehension, lacking in ordinary sense or intelligence, half-witted . . . unlearned, ignorant, easily misled. . . .'

"It seems to me that 'simple' is not a simple word. While I do not argue that simplicity is for simpletons, I believe that it is an irrelevant factor. I want to get things right, whether rightness comes from simplicity or complexity. The history of technology has many examples of designs which were 'right'. Very often, these 'right' designs are elegant."

And so, of course, is the Duck and the literary style of its defender.

***

The CEGB: At the Marchwood Laboratories, Mr. Ian Glendenning, the scientist who first dared to mention the magic figure of 120 GW, moved back to his first love, nuclear energy, and the physicist supporting him, Mr. Brian Count, who is an expert on hydrodynamics, took over. Mr. Count is a steady, quietly-spoken man who gives the impression of being sound and dogged. He is not easily budged, either with other people's enthusiasms or with their despair, and he has just kept on with his work on wave energy while storms have been raging outside. He is responsible for a new device, one which lacks a name and could be called an Air Raft. It uses the concept of the Oscillating Water Column but instead of a deep draught, as in the *Kaimei* and the NEL design, it will be only 4 m or so deep. Mr. Count describes it as "a big box with a hole in the front".

Mr. Count divides the devices into those which depend on mechanical, hydraulic and pneumatic systems..The chief of the mechanicals is the Duck, which remains the most contentious. Mr. Count, in agreement with CEGB thinking, regards it sceptically because, as he put it, "it has an awful lot of mechanical equipment. There is more technology in a square inch than on some of the other systems altogether. If all the mechanical equipment works, it would be fantastic" — a comment expressing admiration as well as doubt.

He has come down against "anything mechanical, which costs a lot of money". He has also turned away from hydraulic systems. "We are worried about them. We are not eliminating them. We believe that there should be a proper study." In this, Mr. Count is echoing the caution expressed by Sir Christopher Cockerell, in one of the earliest interviews I conducted, at a time that seems light-years away, when he expressed his own earthy hesitation about hydraulic systems: "the tiniest bit of grit puts them out". Personally, having once driven motor cars with rod brakes, I have never understood why the car industry went overboard for hydraulics. I have heard of a brand-new Rolls Royce in which the chauffeur pressed on the foot brake and found there was nothing there. The police accept that a pin-hole can develop in a tube and all the fluid will disappear.

One is left with Mr. Count's choice: pneumatic. The waves drive a bubble of air and the air drives an air turbine. Its critics will say that there is a loss of efficiency in taking the direct power of the sea and changing it into air pressure before changing it back into electricity. It also has problems which Mr. Count does not ignore. "It is a matter of debate whether you have enclosed or open systems. Enclosed, you don't get fouling through the turbine. It is better for maintenance. We don't see (as Mr. Salter does) any system being maintenance-free. At least for the first stage, we don't. With an open system (such as the OWC) there could be salting on the turbine plates and icing and we must debate whether that deters from the performance of the turbine. We don't know enough about the problem of water going into an air turbine. It could come from a high wave crashing over the model, or from the water column itself going up into the turbine. You could shear a blade off. But with an enclosed system, you could lose part of your air supply when the bag is punctured.

"The OWC is rated at present at slightly over 10p. There is less speculation in that design and it is a more credible system than some of the others. We have got to balance the costs with the speculation. There is still no clear-cut decision as to which is best. We are a cautious organisation. We have got to go on the cautious line. Our device has no hinge. It is just a big box with a hole in the front, a shallow-draft Raft but really a development of the OWC."

He is an admirer of the OWC concept − "that could be built now".

The OWC: This is the device, invented by Yoshio Masuda, which leads all others in practical development. His ship, the *Kaimei,* was moored off Japan for nine months and officially rated as a 2 MW power station; there are 300 buoys working in the Pacific; and Trinity House by the spring of 1980 had its first three at sea, off Dover, Yarmouth and Harwich. This is the design which Mr. Count identifies as the one which could be built, full-scale, at any time.

It has been the centre of the most encouraging, and the most depressing, developments in wave energy. The Japanese launched, in the autumn of

1978 the *Kaimei,* in the sea off Yura, as a full-scale prototype. It fed electricity into the national grid, floating in moderate waves, with a mean height of 3–4 m, and a maximum of 10–11 m, about 4 km offshore. It sounds a small power station, and it is. In Britain, superior designs are available. The *Kaimei* faced the waves bows on and the NEL design, when it gets off the drawing board, will accept them broadside on. Instead of holes under the ship, it will have open portholes on the side taking a direct buffeting. We shall do better, when we do it.

On the deck of the *Kaimei* was a made-in-Britain generator (Fig. A.2). It was produced by a company in Newton Abbot, Devon, called Centrax.

Figure A.2.   The British-made generator on the deck of the *Kaimei,* Japan's floating
2MW wave power station

It cost £300,000 and was regarded as a British contribution to the International Energy Agency programme. In reality, it was British aid to Japan.

Britain and the US had been invited to contribute, and the American offering never turned up. It encountered what have been called "bureaucratic delays at the Department of Energy" in Washington. Britain's generator was commissioned and built within six months by Centrax. Nothing like it had ever been wanted before. It is an air turbine which spins at up to 1,800 revolutions per minute, producing between 100 kW and 170 kW, depending on the force of the small waves in the Sea of Japan, along with seven other generators.

The British device was supervised by Mr. Len Bedford of the Wave Energy Steering Committee. It was built largely in the open air during the appalling weather which hit the West Country with particular severity in the winter of 1978/79 and was completed with five weeks to spare for tests. It was then taken, at three miles per hour, to Southampton to be shipped out to Japan. It is 20 ft high and, fittingly, had to be accompanied by SW Electricity Board workers with hooked poles to lift up overhead power cables to make way for it. It arrived in Japan, two days before the agreed date, where Mr. Bedford supervised its installation. It was a superb product of British engineering industry, something to mention next time the *Wall Street Journal* and the *New York Times* lament over our decline.

Now for the depressing aspect of the OWC. The parsimonious Government programme set it back badly. There had been a plan (page 90) to build a 1/10th scale model and float it in the mouth of the Clyde in the summer of 1979. It was aborted because of a 10% cut across the board in all NEL work. The British version would surely have been more efficient than the Japanese but as it did not get built we are left with a sad story of might-have-been. The NEL engineers have been demoralised by the way that they have been treated. Some of them drifted away to other activities. Those who remained continued with their studies, on paper or in bathtubs. They reached a number of interesting conclusions.

They progressed from their first design of a moored floating structure to a tethered structure, with perpendicular moorings, and then to a fixed platform, sitting on top of a pylon drilled into the seabed. Finally, they decided that, taking into account the advantages and disadvantages of a fixed device, it would be better to tackle the problem by building what they called a "deep breakwater structure" or a bottom-fixed model sitting on the seabed in 15–20 m of water close to shore. This will be visible from the coast in clear weather – it will be only about 5 km out at sea – but we are talking not about Eastbourne but about the Outer Hebrides, where the economy is not dependent on hotels with a room with a view for tourists. The NEL, which is painstakingly conservative in its claims, estimates that it will produce 5–10 kW/m of device and can contribute 3–4 GW installed capacity along the north and west coasts of Scotland alone. If all the suitable sites around the British coast were used, this would rise to 6–7 GW. "This would represent a significant contribution to the future demands for power, equivalent to approximately half of the projected nuclear capacity in 1990", says an official release from the NEL. The cost was estimated in 1978 as 20p–40p per kWh; in 1979 at 10p–20p; by 1980, this had dropped to 5p–15p. The plan was to construct the base and lower walls in an oil platform dry dock, of which five were available in the Firth of Clyde. The structure would be towed out and installed with rock anchors and the power take-off modules dropped on to the base. In a phrase that ought to shake the Department of Energy, the NEL said: "All the work can be carried out using existing techniques".

The Trinity House buoys are called "Wags" – Wave Activated Generators – by the staff who are looking forward to having many more. The economics are striking. The buoys cost £3,000 each. To change a gas cylinder on a conventional buoy costs more than £200 an hour to send out a ship. It takes on average roughly four hours for the ship to reach the buoy and change the cylinder. So, at £800 a trip, the buoy has paid for itself within a few months and the rest is profit. It is not only the money that matters: to change a gas cylinder usually means lowering a man over the side, perching him on a swaying buoy while he unscrews the gas canister and installs a replacement, and as the buoy is marking a sandbank or a wreck the work is dangerous as well as difficult. The men at Trinity House are delighted with their new device and their experience in the open sea is worth more than all the computer printouts that are being analysed so cautiously in the laboratories and offices where people are pursuing a "scientific programme" – to quote that civil service briefing for Mr. Benn.

The Raft: Those who have followed the story of wave energy from the start inevitably feel particular affection for Sir Christopher Cockerell – he was, after all, the first person to propound it as a way of fighting recession by providing useful employment. His Raft started as a seven-pontoon contouring train, came down to three pontoons and eventually was simplified even more by being reduced to two pontoons. Later experiments divided the smaller, front pontoon into two, down the centre, and as they flipped in different phases they did start to resemble Ducks. Some of the Harwell experts believe that a hybrid incorporating parts of different devices may emerge as the winner.

The problem over which the wise men sorrowfully worried their heads was the hinge and the Cockerell team solved it by inventing a steel and rubber bonded joint, rather like a knuckle, instead of the standard hinge with a pin down the middle exposed to the water. This reduced the cost of production by half. It was also decided to settle for hydraulic drive, sucking in sea water as the fluid.

Calculations were made about mass production. The Rafts would be produced in specially-designed shipyards with robots for the interior work. For the famous 2,000 MW power station, it would need 400,000 tons of steel a year for 10 years. This would produce a "housing estate" of 800 Rafts. The cost is estimated at £4 million a Raft at 1980 prices, presumably reducing as serial production developed. So a full-scale power station would cost on this cautious estimate at the most £3.2 billion. Compare this with the £1.5 billion spent by the CEGB on oil-fired power stations which were rapidly becoming too expensive to run and the £1.5 billion demanded by the Highland Regional Council to build a fast-breeder reactor in Caithness which, if experience of the nuclear industry is anything to go by, will prove a huge under-estimate — and then the damned thing probably won't work. Remember too, that the Rafts will start producing electricity from the moment the first one is launched; you do not have to wait for completion, as in a "normal" power station before you start seeing benefits.

The Duck would cost considerably less than the Raft — perhaps £2,000 million for a 2 GW station and other devices might prove cheaper (see below).

The HRS Rectifier: This is the only one of the "first generation" which was dropped by WESC on the grounds that it could not be brought down in price, at least at present. But it has not been abandoned by Mr. Bott and the Crown Agents. In the spring of 1980, Mr. Bott and Mr. Ted Lawrence, Head of Development Services for the Crown Agents, set out for a tour of islands which were desperate for alternative sources of energy: Fiji, Tonga, Samoa, the Seychelles and, of course, Mauritius where it all began. Finance was available from the Overseas Development Agency, the United Nations Development Programme and the World Bank. Mr. Bott was the first to insist on the need for a "passive" device and could be proved right in believing that it would prove more resilient than the power stations which ride the sea. The NEL, as we have seen, is now tending to accept this argument. His idea was a victim, in Britain, of the obsession with costing the unknown but other countries could well overtake the leisurely pace of development in the oil-rich state of Britain. They have less time to play with.

***

The second generation: As the programme developed, other devices entered the scene. For a while, it seemed as though you were not complete unless you had a colour TV, a deep freeze, a second car and a wave power generator. This is not said to write off the new inventions but it does undoubtedly baffle the public when it is faced with such a profusion of ideas and such a poverty of application.

The Lancaster Airbag was first to emerge. Indeed, it was early enough to be included in the Heathrow Hotel conference, though with the proviso by its inventor, Professor Michael French of Lancaster University, that there was a "critical area of uncertainty". But it made an invaluable contribution to the credibility of the programme because it was — with reservations — accepted by RPT at that time as the only one which could produce electricity for 5p–10p a unit.

Professor French is a mechanical engineer who has worked on nuclear energy and who has never weakened in his insistence that wave energy is no more than "insurance technology" for use if "the price of fossil fuels rises considerably or the development of nuclear power is restricted by public opinion". I may have annoyed him by pointing out in *New Civil Engineer* (July 12, 1979) that both these things were happening. He wrote to me, before the first edition of this book had appeared, saying "it is a pity your book takes the line it does".

He explained that he considered wave power "not practical as yet, a view I still hold". Nevertheless, he received me courteously and with his colleague, Dr. Robert Chaplin, explained his approach.

He came to wave energy after working on harbour defences: he devised a scheme for positioning flexible Airbags on the seabed at the entrance to harbours to provide what he called "bubble breakwaters" to calm down the water. That was in 1960. Then, in 1974, he read a paper by Stephen Salter and became convinced that "there was not enough energy in the waves to make it worthwhile, and I still think it is possibly true. The figures were obtained from an untypically favourable situation, Station India." Professor French has had the satisfaction of seeing the potential scaled down and his analysis to that extent justified.

His plan is to have a beam of concrete with the bow facing the waves and a flexible rubber bag attached underneath it. As the water rises and falls, different compartments expand and contract, driving air through a turbine. The advantage is that the flotsam has no chance of being swept into the blades of the turbine; the disadvantage is that when a bag is punctured, the air escapes. With the OWC, the air comes from the atmosphere and damage to the structure would be less serious. Professor French originally planned a long beam with the flexible bag on top. He developed this into the latest design, in which the bag is underneath and the valves are above the water level. It will have protective devices to isolate any bag that gets punctured.

He told me: "There is a lot of experience of this kind of rubber in sea water. Unlike with car tyres, you are not rolling it down on sharp objectives. It is like the side wall of a tyre, which is flexible, and the sharpness of flexing is less than on the side walls of a tyre."

He accepted that, in tests in Salter's tank, the device had shown "peculiar behaviour" in a mono-directional sea: it swung at an angle, instead of facing the waves head-on. But this, he said, would not happen frequently in a real sea.

He concentrated on what he saw as "the really expensive thing in wave power — the structure" and he brought it down, leapfrogging other, longer-established schemes. It is about one-quarter the size of the Raft or Duck for the same power, in the view of its supporters. It has a big piston area compared with the total device area. A generator 200 m long would have an installed capacity of 8 MW and a mean output of 3–4 MW.

Professor French combines his pride in his own device with a continuing restraint of enthusiasm for wave energy and it is impossible not to admire his doggedness in sticking to his tack, even when he is one of the inventors.

The Lanchester Clam: The Lanchester Polytechnic is in Coventry and has become accustomed to its visitors turning up at Professor French's headquarters in Lancaster. The Poly, as we have noted, carried out controversial trials on Loch Ness on its own variation of the Duck but is more closely related to the Airbag. As such, it is an example of the way in which hybrids are developing, taking in some aspects of several designs. The Clam is an Airbag surrounded by a box with a hinged flap. But it lies broadside on to the waves and the hinged area nods as the waves rise and fall, squeezing an air container. The advantage is that the outside flap protects the bag from the sea and its detritus. The disadvantage is that it costs more — in the Spring of 1980 it was listed at 6p—7p a unit. Output ranges from 7.5 kW/m mean to 35 kW/m peak.

The bag would be made of corded rubber fabric, again rather like car tyres and the plan is to use emergency sealing to isolate any bag which became punctured. The thinking behind it is that it would "avoid rather than solve some of the problems", to quote the official description. In particular, it will have a mechanism to enable the flap to shut up completely in a high sea. The means of operating the shutdown had not been devised at the time that I spoke to Mr. Brian Loughridge, the project coordinator at the Poly, though Mr. Salter's adaptation of the gyroscope to wave power would seem to be an obvious option.

The Bristol Cylinder: This is the invention furthest removed in concept from the traditional idea of capturing the waves and it enjoys a great deal of respect from the other teams. It consists of a huge concrete dustbin lying on its side three metres below the surface. In this way, it avoids being buffeted by the weather at the roughest point — where the wind meets the sea (I refuse to use the word "interface"). The cylinder, 50 m X 12 m diameter, rotates in what engineers call "an eccentric orbit — an irrotational motion". That is to say, it moves rather like a bottle floating in the sea, but in the underspin of the waves, below the surface. It follows the same path as the particles of water — an up-and-down, rolling motion with brackets on the side staying horizontal.

It is held down by chains which are tied to hydraulic pumps sitting on the seabed. The cylinder's motion causes the pumps to rise and then, as the chain slackens, the pumps are pulled back into their jackets by springs. The effect of this is to drive sea water through pipes at high pressure and thus to a turbo-generator producing electricity. There was some debate about whether the electrical equipment would also be on the seabed, out of the weather, or perhaps on a platform above sea-level. A line of Cylinders 40 km long would be needed for the 2,000 MW power station.

The advantage, plainly, is that the Cylinder itself and part, and perhaps all, of the power station would be sheltered under the surface of the sea;

the disadvantage comes with maintenance and servicing below the water.

It has been designed by Dr. David Evans, Reader in Applied Mathematics at Bristol University, in association with McAlpine's whose contribution is supervised by Dr. T. L. Shaw, a civil engineer also from Bristol University. One would be tempted to dismiss it as a submariner's dreamchild, were it not for the assessments that have been made by RPT, WESC and, perhaps surprisingly, Stephen Salter who, as we have seen, does not shrink from complexity.

It is the one device with what Dr. Evans calls "two degrees of freedom − horizontal and vertical motions combined" (up-and-down and rolling). It is also the one invention in which theory came first. A Professor T. F. Ogilvie of Ann Arbor Department of Naval Architecture in Michigan, discovered that if you take a cylinder in a narrow tank, like a rolling pin in a bath, and force it to rotate below the surface, waves are generated in one direction only. It seemed to be simply a scientific paper, buried in abstruse mathematics in the *Journal of Fluid Mechanics.* That was in 1963. Today, money spent on pure research with no apparent practical application is justifying itself.

There are many other designs being developed. One of the most important is the Belfast Buoy, which sounds like an Irish folksong but is actually a relative of the OWC. It has a round shape and can accept energy from any direction. On top, it has the Wells self-rectifying turbine, which can accept air from below and above. It is better than anything possessed by the Japanese who have bought an option on using it on their buoys.

There is no shortage of ideas, no shortage of talent, no shortage of enthusiasm.

CHAPTER 11

Conclusions

We know now that the waves contain the power that we need. We know that the technology exists to tap that power. We know that there will be an endlessly growing demand. Yet we appear to be in a situation of hiatus. Committees meet, computers are programmed, research goes ahead at a gentle pace. The feeling of excitement and urgency that are to be found among, particularly, Salter's group and to a lesser extent the Cockerell team are in no way matched by the attitude of the Department of Energy or the CEGB. Indeed, there are grounds for feeling that the two bodies which have most to gain from the development of new sources of energy are being deliberately unenthusiastic. Caution, one could understand; but the official attitudes seem less than that.

Consider, for instance, the White Paper published by the Government on June 6, 1978 — a major statement of policy on the subject, entitled The Development of Alternative Source of Energy. It was produced four months before the expiry of the Government grant of £2.5 million for research and development into wave energy and it offered a further £2.9 million and promised that the programme would be reviewed annually from the spring of 1979. Consider what this meant on the ground: no-one could plan beyond the next spring.

There was one phrase in the White Paper which shocked even some people who were the least critical of the Government's progress: "The programme could build up substantially over the next three or four years to a level of expenditure that will enable a single device to be identified on which resources could then be concentrated." This would be a death blow to at least 80% of the people engaged in wave energy. No-one that I have interviewed has ever before suggested that by 1982 or 1983 it would be possible or right to shut off all help from all but one of the groups of experimenters. Never in history has a major new technical development been introduced without improvements and changes from other enterprises. Supposing we had put all our energy development into Zeta, and ignored the possibility of development of nuclear power, where today would the Department of Energy be?

But surely, the reader must be thinking, this is obvious to lots of other people. Surely there are civil servants who realise this, too? Surely indeed there are. And what they are spelling out in this White Paper, to which the Government lent its authority, is that they were not taking wave power seriously. We were being treated to an exercise designed to defuse the environmentalist argument. There is, in the view of some of the people closest to the situation, no real danger from nuclear power. Coal will last for 400 years. North Sea oil will last twice as long as forecast (the oil companies always under-estimate the size of a field, usually halving the expected total; this is partly from caution but it does help shares when they announce that the wealth is greater than expected). But then there are problems with chaps who don't like pollution. Look how nasty things have got in Japan and France and West Germany. We don't want a Green Party here. So let's throw them a bone. Let's put out a White Paper, well before energy might become an election issue, and shut them up.

The tactic can be seen even more clearly in the White Paper's approach to tidal power, which means primarily and almost entirely a Severn Barrage. It is a hot political potato. Mr. Benn and Mr. Arthur Palmer, then influential chairman of the Select Committee on Science and Technology, are Bristol MPs and anxious not to shut the door on the project; equally, they were reluctant to go ahead because there were local interests, ranging from salmon fishers to hotel keepers, who were unenthusiastic. So a new committee was set up to make "further studies" and it was given £1.5 million to pay for its labours. And the Energy Department's Chief Scientist, Sir Hermann Bondi, was the chairman. He has never built a barrage in his life. But what the Government wanted was a time-consuming device that would take the heat out of the issue. It is a Bristol issue, just like Concorde was, and it must not be allowed to become a political issue.

If this sounds cynical, consider a report in *The Times* of April 11, 1978. Its Science Editor, Mr. Pearce Wright, said that a committee report circulating in the CEGB "suggests it is important to explore alternative (sources) both to satisfy the board that nuclear expansion is fully justified, and to demonstrate that to other groups opposing nuclear power expansion".

Yet it does look as though the CEGB may have stumbled into a nasty case of indecent exposure. It set up research at Marchwood and may well have been hoping that the numerous difficulties that would be revealed — and which have been revealed — would help to steer opinion away from these new-fangled ideas. Mr. Glendenning's studies have certainly underlined the difficulties, as any scientist or engineer would have had to do. But he has also come up with the figure of 120 GW, and no matter how the critics may seek to emphasise the genuine problems, they are left with the hard fact that a major source of power is waiting in the sea to be tapped. Mr. Glyn England has now accepted the Glendenning conclusions. A cynic might note that he did so at a time when the Department of Energy was trying to foist on the CEGB the expansion of its coal-fired power stations, which are expensive to run, compared with nuclear. So Mr. England may have been encouraged by that fact to gaze into the deep blue sea. But even if that had been an element in CEGB thinking, we are still left with the statement by Mr. England that the waves could "supply the whole of Britain with electricity at the present rate of consumption." Who are its opponents?

At the heart of the opposition are the nuclear scientists, who have spent a lifetime studying a 20th-century discovery of major importance and who could now face the prospect of being relegated to a less significant role, while technologies in which they are not expert become a more attractive alternative. The nuclear lobbies are powerful in the Department of Energy and at the CEGB. In both centres, you will hear endlessly of the convenience of pressing a button and bringing a nuclear power station on stream. It was not from any of these people that I learned that nuclear power cannot become the main, firm source without extensive development of storage facilities.

It is heresy to suggest inside the Department's offices that we could obtain all the electricity we need from the waves. It is, of course, a provocative statement to make to anyone but inside the Department it produces an electric response. "Say that again", one of their top people said to me menacingly.

Some of the environmentalists are not helping their case. They have magnified the message that Small is Beautiful. One almost feels that if only someone could produce a nuclear power station on a silicon chip for every home, they would embrace nuclear power. They want small power stations, even if they burn coal and make smoke in every town. They want fluidised beds, regardless of smoke and sludge, so long as they are small. Wave energy generators can be big and clean – but big.

Then there are other rival interests to be considered. Dr. Freddie Clarke, Research Director (Energy) at Harwell is the key figure in our government's thinking about future needs. His attitude is that alternative sources are "likely to make only a minor contribution" this century but that there could be "a take-off in their use during the first decades (plural) of the next century". He made the statement while addressing the National Federation of Women's Institutes and he used a form of argument which is common among the people contemplating wave energy.

He sought to blind his listeners with engineering. The structures involved, he said, would be the size of aircraft hangers. One element of any of the devices could contain the hall they were meeting in – Westminster Hall. The ladies left suitably put down.

This form of argument is difficult for the layman. I have been told about buildings as high as a block of flats and as long as a giant oil tanker, pounded by waves strong enough to pick up a cathedral and throw it around a circle of 30 m diameter. When the speaker is a distinguished engineer or scientist, it is difficult to contest his case. But let us try to.

"Imagine trying to build a wall 365 m above sea level, containing 100,000 m^3 of concrete. Just for a start, old boy, how do you get the concrete up the mountainside? It would have to contain 10 million cubic metres of water and to get the water up to that height would require pumps weighing about 650 tonnes inside a cavern dug out of the earth, 82 m long and 30 m high, big enough to take a seven-storey building . . . ." The mind is suitably impressed. It sounds like an impossible project. Yet it is in fact a description of the Cruachan pumped storage development in Scotland, which has been functioning for 13 years with a 400 MW capacity.

Dr. Clarke is, with justification, a great admirer of the achievements of the oil companies. And *they* have no wish to see any rival source of energy which might make their presence less desirable.

Then there is the coal lobby. Mr. Alex Eadie, who was No. 2 in the Department of Energy, represents Midlothian and was chairman of the

Parliamentary Labour Party's miners' group. The National Coal Board, with the support of the NUM, is digging up the Vale of York around Selby and the assurances given at the public inquiry about how little damage it would do to the area have not worked out; local people feel that they have been misled. The NCB is now contemplating the Vale of Belvoir. It also has plans for the Cotswolds and there are rumours that the city of Oxford is sitting on a pillar of coal. These plans could one day provide tens of thousands of jobs (and of NUM members) if society decides that it needs to spoliate the countryside and condemn an army of men (and perhaps women) to spend their working lives underground in uncivilised conditions. (One of the most prominent of the miners' leaders, Mr. Arthur Scargill, emerged as an environmentalist in opposition to the extension of nuclear power during the Windscale inquiry; he at least has continued courageously to pursue the same course when the consequences will be less rather than more dependence on coal).

None of these factors should be exaggerated. We are talking about honourable people, genuinely concerned with our future and worried, in the case of the civil servants, about recommending the commitment of enormous sums of money to an experimental project. Yet their problem is not eased by the pressures applied to them by people dedicated to rival interests.

Consider in series the objections that have been raised to embarking on a programme designed to put full-scale prototypes, or parts of them, into the open sea as fast as they can be produced, giving the order (and the money) to go ahead immediately:

(1) "We would make mistakes." Yes, of course, and the most trenchant answer to that one has come from Cockerell, who risked his own money before the Government gave help to his scheme. "In the development of anything, if it is done well it must show a waste of money. But the Civil Service can't justify it. It has not got a column to justify it" he told me. "But the designers have got to have a chance of making mistakes. We know we can produce electricity." To which can be added the view that until these devices are at work in the open sea, all the testing tanks in the world and all the computer studies will also be wasteful because the sea always has a trick up its sleeve. Nothing can equal the real ocean.

(2) "The first ones will be inefficient." This is the weakest argument of all. The best answer has been provided by Salter: "Efficiency itself is of no concern when the gods pay for the waves." And, in the words of Cockerell, "we don't have to standardize at any time. With the railway system, you have

to decide a gauge and whatever forever. With wave energy, you could have a fleet of devices which could be one design, and you could have a Mark 2 next to it and a Mark 3 next to that. That is one of the delightful things about it." Then there is Glendenning's point that because the devices are "modular", you can reproduce them and improve them, unlike a tidal barrage scheme which demands a huge sum of money as a once-off payment.

(3) "The waves will never represent a major contribution to our energy needs, or at any rate not until after 2020 or maybe 2030." This would be true if the Government was not shifted from its present cautious approach. But we know that there are 120 GW around our shores and no-one has yet provided a convincing reason why we should not contemplate winning a large part of that. If we pursued a programme with the vigour that the Japanese are showing, then we could be building 1,500 MW power stations at sea within two years.

So what sort of programme should we be thinking about?

First of all, the most important development in wave energy has happened almost by chance, almost unnoticed. Wave power now exists at a huge range of sizes, greater than any other existing alternative. It can be used to light a 60 watt bulb and drive a flasher unit on a navigation buoy based on the OWC. It can be used for a single Salter Duck with an output of 1 MW, costing about £1 million. It can be used for one Raft producing 2.5 MW and costing about £4 million. It is used for a ship like the *Kaimei* which would be rather more expensive if it were a long-term fixture sitting on the seabed, as the NEL intends, for capital costs, but would be more fruitful. Then there are reasons for believing that some of the "second generation" devices, particularly the Airbag, the Clam and the Bristol Cylinder, might reduce costs still further, once they have been fully tested.

All of them, it must be remembered, would produce electricity from the moment that the first unit was launched. Contrast this with a standard power station which starts to be useful only when the whole edifice has been completed. In the case of a nuclear power station, it takes 12 years before any electricity flows and it costs at least £1 billion. During those 12 years, wave power generators would be defraying their expense by sending out electricity. And always all of them would offer the promise of growing into a full-scale, 2,000 MW power station.

But where, one might ask, could we make a useful contribution NOW? The answer has come happily in the spring of 1980. The very area which has the most fruitful seas also has the most expensive electricity in Britain. The Western Isles are in trouble. They are lit by diesel generators. The islanders are given a subsidy, paid by the mainland consumers. The real, unsubsidised cost of diesel-electricity is between 25p and 27p a unit. This is the figure which should be compared with the cost of wave electricity, which can be provided by four devices for less than 10p a unit.

It would be wrong to make too much of this. The supporters of wave power would not wish to make the same blunder as was made by the electricity industry in the 1950s when it was promising cheap (and even free) electricity from the great new nuclear invention. In the 1960s, the oil companies waged a great advertising campaign urging everyone to switch to oil-fired central heating and it depicted "Mrs. 1970" in this way:

"As to future fuel bills, oil has the lowest running costs. What's more, oil-fired central heating puts up the value of your house."

Lady Isobel Barnett, the radio personality, was featured as the sensible housewife who was delighted with the switch to oil. Before the 1970s were out, she had changed to high-speed gas. The moral is that boasting about costs, even for gas which in 1980 was having its price artificially increased by Government policy, is liable to prove misleading. This applies particularly when long-term construction is involved: Dinorwic pumped storage station was estimated to cost £80 million in 1974 and this had risen to £460 million within six years and was expected to cost finally £500 million.

For wave electricity, one would not wish to repeat these spectacular blunders. But it is undeniable that the waves could supply the Western Isles with electricity for considerably less than the diesel generators and it is a fact that the cost of oil is certain to increase while the cost of capturing the waves is certain to come down. And the fossil fuels which cost more are finite, while the waves will last for ever. The Highlands and Islands Development Board carried out a survey of the possibility of using the waves for the islanders and concluded that a 20 MW wave power station (a 50% increase on existing demand) could be absorbed by the islands "within a few years". The capital cost would be £20–£30 million for a line of Ducks. The cost would be defrayed every time someone switched on a light.

It would not need much political courage to recognise that the Western Isles were an ideal site for a seaborne test bed. The problem which arises is which device to choose. If it were only for one small, once-off, 20 MW power station, it might be of transient consequence. But a courageous Secretary of State for Energy should, in my opinion, act as though there was a war on. He should imagine what action he would be taking today if, for instance, the Middle East was suffering a nuclear attack or more simply deciding for political reasons not to make its oil available to the West, and the Common Market countries were demanding a share of our oil and coal. The energy famine, which is sure to come sooner or later, had already arrived.

The tendency would be to choose what is being called a hybrid – that is to say, a mongrel. The Raft is looking more like the Duck, the CEGB is combining the OWC and the Raft, the Belfast Buoy is similar to the OWC. The Airbag and the Clam are closely related, so why can't we combine the best qualities of several inventions? I have one of my own to add. It is called the Neptune, after the Greek sea god. There will be a Raft as the basic platform. A spine will be stretched across its stern to support a line of Ducks. A hole in the Raft will open the way to an Oscillating Water Column. There will be vents in the side admitting water, which will rush through a turbine producing an HRS Rectifier. On the way, it will pass through a pond providing a fish farm. Under the Raft, an Airbag will be strapped on. Chains will go down to the seabed, driving pistons. On deck, there will be room for an aero-generator capturing the wind, and a solar panel.

At this point the ultimate pessimist, from the CEGB no doubt, will ask what happens on an overcast day, with little wind and no sea running. The answer is that we will ring up Dungeness and ask if they happened perchance to be functioning once again.

This is intended to illustrate a serious, growing problem. With the proliferation of devices, and the healthy comparison of ideas, there is a temptation to fudge options. Either the Government will settle on one device and press the abort button for all the others, or it will play around with combinations and carry on talking, talking, talking.

A preferable policy would be to recognise that we need to get into the sea quickly. The determined Minister would do as Beaverbrook did with the aircraft manufacturers in wartime. He would call together Salter, Cockerell, Bott, Russell, experts from the NEL and IOS, Harwell and Marchwood, and tell them that he wanted results immediately. Money was less precious than time.

It would be accepted that "mistakes" – in the sense that some devices would prove superior to others – would be made, but that we needed to start producing electricity right away. With or without a "low energy strategy", no-one doubted that eventually we would need the waves.

From such a conference, a serious programme would emerge. We could launch Ducks, Rafts and OWCs within months and their bounty could easily be absorbed, with a saving in cost, by the Western Isles. Or we could adopt the scheme put forward by the Cockerell team at the Heathrow Hotel conference of developing pre-production units smaller than what they call "the Hebridean scale" for less turbulent seas. Their proposal was to use the S.W. Approaches or the North Sea for a floating test-bed "for both detail design and line production methods, provided it is large enough to involve all the appropriate technology." But since then a new indoor testing tank for toy boats has been opened at Cockerell's headquarters, for use on devices that are two-to-three years behind the Raft and the Duck, and the day when the real sea is confronted has been put back.

The alternative plan would lead to a continuing programme of production. Expenditure of £1,000 million or even £200—£250 million could transform the energy scene in a way that could not be done by investing that amount in any other source of electricity – conventional, nuclear or renewable.

At the least it would enable us to slow down from the criminal rate at which we are burning our finite fuels. It would also provide us with products with an almost inexhaustible export market. And if – or rather, when – the political realities of the Middle East impose themselves on a world desperate for energy, there would be a safe alternative available.

At home, it would provide a means of reviving the shipbuilding, steel and construction industries. The shipyards and the builders, normally dependent on once-off orders for a single job, would be delighted at the prospect of continuing production of identical modules. The steel industry would leap at the opportunity. Soundings have been made in all three areas.

What is needed is the political will. The alternative is a policy of contraction, with 2,000,000 unemployed, shipyards dying and skilled workers rotting in the very areas where the economy could be given a tremendous surge with orders for something productive, creative and fruitful – wave energy generators, the renewable, benign sources of power which can take us out of the polluting age when we burned coal, oil and uranium, and into a revolution in technology.

Appendix

Wave Language

The key factor in assessing the energy in a wave is its height, just as most of us would imagine without any technical knowledge. It is the high wave which lifts us up as we swim and throws us the greatest distance. But there are other factors of significance. These are some of them. They are drawn mainly from the studies of the Institute of Oceanographic Sciences and the Hydraulics Research Station.

The *height* of a wave is the distance from the trough to the crest, not to the height above sea level. As the waves rise and fall, the volume of water above sea level is roughly equal to the volume of water below. The sea does actually rise and fall, just as it seems to be doing when we are on a ship.

The *Significant Wave Height* is a term used for convenience which refers to the highest one-third of the waves. It is obtained from a chart drawn on a moving drum of paper and looking rather like an unhealthy electro-cardiogram reading. The pen receives messages from a wave recorder. The earthy aspect of the science is that it has been discovered from practice that the SWH is almost identical to the ancient mariner's tale. When an experienced seaman reports on the size of the waves that he has survived, the figure he gives has been found invariably to be close to the highest one-third of the waves.

The *fetch* is the distance in nautical miles over which the wind has blown continuously on the sea before reaching the point under consideration — in our case, the line of wave energy generators. In general, the longer the fetch, the higher the waves. Strong winds of short duration can whip up higher waves than gentler winds blowing for a longer time; against this, a weaker wind over a greater distance can create very high waves. But it is still the length of the fetch which is the biggest factor.

The fetch is broken by any strip of land on which the waves break.

Then a new fetch has to be calculated from zero. The fetch will also be broken by a line of generators which will absorb the energy and create a calm patch of water. There is some dispute about how much distance must be left between parallel lines of generators to enable the waves to grow significantly again. Many people believe that there will have to be a gap of 160 km but there is a growing feeling that a 80 km gap, followed by generators half the size contemplated, would make a significant contribution.

The *Zero Crossing Point* is an imaginary dot on a dead calm sea, something which we normally call sea level. It is roughly half-way between a crest and a trough. It is also used, horizontally, to measure the time and distance between waves.

Wave frequency is a term familiar to anyone associated with radio or, indeed, any form of noise. But it is best to avoid its use when discussing sea waves because they, unlike air waves, do not travel at the same speed. For instance, the longer the *wavelength*, that is the distance between the lines of crests, the faster the waves travel. A better way to describe the waves is to consider their *period*, that is the time between waves. A typical time is around eight seconds. For those who feel more at home with frequencies, it may be helpful to note that the frequency is the inverse of the period. Thus, an eight-second wave has a frequency of one-eighth of a cycle per second.

The *power* in a wave can be expressed by the formula

$$P = 0.55 \, H_s^2 \, T_z \, kW/m$$

It is not as formidable as it looks. It means that power (P) equals 0.55 times the square of the Significant Wave Height (in metres) multiplied by the Zero Crossing Period (in seconds). The result is in kilowatts per metre. Let us take an example.

Imagine a wave of 3 m height with a Zero Crossing Period of 6 seconds. Then the power equals 0.55 multiplied by 3^2 multiplied by 6. That is, 0.55 x 9 x 6. The answer is 29.70 kW/m of wave front. And, given that we are dealing with a number of unknowns, it is safe to simplify the formula: instead of 0.55, one can say "a half". The answer, in the example given, would then be 27 kW/m. Until wave energy devices are operating on a large scale, the difference is not important.

What is important is to bear in mind that the height is squared. So height is more important than period. In the example given, a wave of only one metre more in height would produce the equation $0.55 \times 4^2 \times 6$, which would produce 52.80 kW/m. One metre more in the wave's height, the size of a child, and its power would be nearly doubled.

Another formula, enabling the wavelength to be calculated, is $L = 5.12T^2$, where T is the time in seconds. This is useful where instruments exist to measure the period between waves, while the wavelength is more difficult to measure. The wavelength is an interesting figure as a description of a sea, particularly for those who go out in small boats. It is less significant for those dealing with wave energy, except for the Cockerell Raft; for most inventors, the height and period are the more important figures. For those who like to have a mental image of the sort of conditions we are discussing, under the formula quoted an 8-second wave would have a wavelength of 100 m. A 9-second wave would have a wavelength of 131 m.

The waves are created by the wind. Its energy is transferred to the water at the surface by a method which is still not fully understood. But there is agreement among experts that the waves represent a concentrated form of wind energy and that is one reason why wave energy has overtaken aero-generators (giant windmills with horizontal arms), even if the latter are stationed in the sea.

A generally-accepted guide to the power in the waves is that they produce roughly 70 kW/m off the Outer Hebrides, 25 kW/m in the Irish Sea and off the north-east coast of England, and 50 kW/m off the Scilly Isles. As a convenient figure, it is usually the figure of 50 kW/m which is taken as the average and this gives 50 MW/km — kilowatts to the metre and megawatts to the kilometre.

Selected Bibliography

The Development of Wave Power – *a Techno-economic Study*, by J. M. Leishman and G. Scobie, Department of Industry and National Engineering Laboratory, £5, 1975.

Power Plus Protein from the Sea, by Walton Bott, *Royal Society of Arts Journal*, 40p, July 1975.

Energy From Waves, by Ian Glendenning, CEGB, Paper presented to Oceanology International, March 1978.

Energy from the Sea, by Ian Glendenning, *Chemistry and Industry*, July 16, 1977.

Wave Energy press releases, Energy Technology Support Unit, April 1976 and 1977.

The Development of Alternative Sources of Energy, White Paper, Department of Energy, June 1978.

Light from the Face of the Deep? Stephen Salter, University of Edinburgh, 1974.

Wave Power – Nodding Duck Wave Energy Extractors, Stephen Salter *et al.*, Energy from the Oceans Conference at N. Carolina State University, Jan. 1976.

The Architecture of Nodding Duck Wave Power Generators, Stephen Salter *et al.*, *Naval Architect*, Jan. 1976.

Wave Power, Stephen Salter in *Nature*, June 1974.

Power from Wave Energy by Walton Bott, *Crown Agents Quarterly Review*, Winter 1976/77.

Energy from Sea Waves, Wavepower Ltd.

HRS Notes, Hydraulics Research Station.

Research and Development of Wave Power Electricity Generation System with Wave Breaking Function, Japan Marine Science and Technology Center.

142

Status Report on the Alternative Energy Sources, Dr. F. J. P. Clarke, Research Director (Energy) Harwell.

Waves a Million, *New Scientist*, May 6, 1976.

Energy from the Oceans, Dr. Walter Marshall, then Chief Scientist of the Department of Energy.

Royal Commission on Environmental Pollution, July 1975, Evidence by Department of Energy.

Alternative Energy Sources for the UK, Dr. J. K. Dawson, head of the Energy Technology Support Unit.

Offshore Technology, National Engineering Laboratory.

Ocean Wave Power, by Ian Glendenning, Central Electricity Generating Board.

Wave Power Availability in the NE Atlantic, *Nature*, September 1976.

Windmills and Watermills, by John Reynolds, edited by David Braithwaite, Hugh Evelyn, publishers.

Industrial Archeology of Watermills and Water power, Heinemann Educational Books/Schools Council by Geoffrey Starmer and a group of contributors.

The Wealth of Some Nations, by Malcolm Caldwell, Zed Press Ltd.

The Energy Question, by Gerald Foley, Penguin Books.

When all the Oil is Gone, by David Ross, *Sunday Express*, Sep. 25, 1977.

Will Britain Miss Out on Wave Power, by David Ross, *New Statesman*, June 9, 1978.

An Energy Revolution, by David Ross, Observer Foreign New Service.

Wave Energy, ETSU, Harwell (Energy Paper 42), HMSO.

Wave Energy Conference at Heathrow Airport, Proceedings, ETSU, Harwell, HMSO.

Index

145